COMPACT SECOND EDITION
The Mouse Brain in Stereotaxic Coordinates

COMPACT SECOND EDITION
The Mouse Brain in Stereotaxic Coordinates

George Paxinos
Prince of Wales Medical Research Institute
The University of New South Wales
Sydney, Australia
(g.paxinos@unsw.edu.au)
(http://www.powmri.edu.au)

Keith B. J. Franklin
Department of Psychology
McGill University
Montreal, Quebec
Canada H3A 1B1

ELSEVIER
ACADEMIC
PRESS

Amsterdam   Boston   Heidelberg   London   New York   Oxford
Paris   San Diego   San Francisco   Singapore   Sydney   Tokyo

This book is printed on acid-free paper.

Copyright ©2004, Elsevier Science (USA).

All Rights Reserved.
No part of this publication may be reproduced or transmitted in any form or by any means, electronic or mechanical, including photocopy, recording, or any information storage and retrieval system, without permission in writing from the publisher.

Permissions may be sought directly from Elsevier's Science & Technology Rights Department in Oxford, UK: phone: (+44) 1865 843830, fax: (+44) 1865 853333, e-mail: permissions@elsevier.com.uk. You may also complete your request on-line via the Elsevier Science homepage (http://elsevier.com), by selecting "Customer Support" and then "Obtaining Permissions."

Academic Press
An imprint of Elsevier Science
525 B Street, Suite 1900, San Diego, California 92101-4495, USA
http://www.academicpress.com

Academic Press
84 Theobald's Road, London WC1X 8RR, UK
http://www.academicpress.com

Library of Congress Catalog Card Number: 2003113435

International Standard Book Number: 0-12-547640-X

PRINTED IN THE UNITED STATES OF AMERICA
04 05 06 07 08   9 8 7 6 5 4 3 2 1

Dedicated to
Melpo   Heather

Contents

Preface ix

Acknowledgments ix

Introduction ix

General Methodology ix

Methods x

Histology x

Preparation of Images xi

Nomenclature and Abbreviations xi

Drawings xii

Use in Surgery xii

Use of the David Kopf Stereotaxic Instrument xii

The Brain Blocker xiii

The Basis of Delineation of Structures xiii

Forebrain xiii

Hindbrain xv

References xvi

List of Structures xix

Index of Abbreviations xxiv

Figures 1

PREFACE

The Compact 2nd Edition of The Mouse Brain in Stereotaxic Coordinates was produced to make available the coronal diagrams of the atlas at a reduced size and reduced price.

An atlas is a record of how we see the world and represent it to our contemporaries, and neuro–anatomical atlases are no exception. This book is a window through which readers may view the organization of the mouse brain as is now understood.

The recent great advances in molecular biology provide impetus to using the mouse. Because of its short generation time and the availability of a large number of strains (including knockout strains), it has been the key mammalian species in research on genetics and development. Its rapid growth and short life span are also advantageous for studies on the biology of aging, while the characteristics of the mouse immune system are favorable in transplantation studies.

We decided to construct an atlas of the mouse brain because of the allure of mapping the relatively uncharted brain of this species, and the perception that this book would be of use to the neuroscience community, which has a growing interest in the mouse. Although there have been several atlases of the mouse brain produced, until the first edition of this atlas there had been no new atlas of the whole mouse brain since the mid–1970s. During this time there have been major advances in anatomical knowledge, particularly as a result of neurochemical mapping.

In this atlas we have summarized the current knowledge of brain anatomy and supplemented it with our observations of the mouse. To facilitate comparison between the mouse and the rat, we have retained the general methodology and familiar format of *The Rat Brain in Stereotaxic Coordinates,* 4th ed. (Paxinos and Watson, 1998).

Features of the Compact Second Edition:

1. It is based on the flat–skull position, with bregma, lambda, and midpoint of the interaural line usable as reference points.
2. A coronal set of 100 diagrams delineating the entire brain.

Reproduction of Figures by Users of the Atlas

For permission to reproduce figures contained in this publication, please contact the publisher at the following address:

Academic Press
Permissions Department
6277 Sea Harbor Drive
Orlando, Florida 32887
Telephone: 407-435-3990
Fax: 407-352-8860

Permission to reproduce a limited set of figures is usually a routine matter. Please identify the figures you wish to use and allow approximately 4 weeks for your request to be processed. The same procedure should be followed for reproduction of figures from *The Rat Brain in Stereotaxic Coordinates*, 4th ed. (Paxinos and Watson, 1998), *Atlas of the Developing Rat Nervous System*, 2nd ed. (Paxinos et al., 1995), *Chemoarchitectonic Atlas ofs the Rat Forebrain* (Paxinos et al., 1999b), *Chemoarchitectonic Atlas of the Rat Brainstem* (Paxinos et al., 1999a), *The Rhesus Monkey Brain in Stereotaxic Coordinates* (Paxinos et al., 2000) *Atlas of the Human Brainstem* (Paxinos and Huang, 1995), *Atlas of the Human Brain* (Mai, Assheuer and Paxinos, 1997), and *The Chick Brain in Stereotaxic Coordinates* (Puelles et al., in press).

The authors would like users of the atlas to consider the suitability of the nomenclature and abbreviation scheme in this book for their own work. The scheme used here is systematic and is based on the principles of the construction of abbreviations for the elements of the periodic table and abbreviations of words such as acetylcholinesterase (AChE).

INTRODUCTION

General Methodology

The brain depicted in this coronal atlas Figs 1–93 was the most complete in a sample of brains from 26 adult C57BL/J6 mice (weight range 26–30g). The caudal medulla (diagrams 94–100) were obtained from another mouse of the same strain and weight. The choice of mouse strain and weight range presented some difficulties because mouse strains vary much more in weight and brain anatomy than rat strains (see Wahlsten *et al.*, 1972). The C57BL/J6 strain was chosen because it is one of the most widely used strains and is of intermediate size. The weight range selected represents the fully adult male mouse. Stereotaxic surgery on the mouse is relatively difficult because of the great fragility of the mouse skull. We therefore chose a weight range representative of the fully adult mouse in which the skull is calcified and presumably stronger.

Some adjustments for use of this atlas with other strains can be calculated from Wahlsten *et al.* (1972) who show stereotaxic grids in the sagittal plane for seven strains of 77– to 78–day–old mice, with the same skull flat orientation as this atlas.

Plates are usually 120mm apart, except where no usable sections were available, and the interplate distance is as small as 40mm or as large as 160mm. Alternate sections are stained with cresyl violet or are reacted to reveal acetylcholinesterase (AChE). Intermediate sections were processed immunohistochemically to reveal parvalbumin. These sections were used to aid delineation, but are not reproduced in the atlas. Other histochemical markers used to assist delineation, but not shown in the atlas, include immunohistochemical stains for substance P and tyrosine hydroxylase, Timm's reaction for zinc, the NADPH–diaphorase stain for nitric oxide synthase, and the iron–hematoxylin stain for myelin.

Structures were delineated using *The Rat Brain in Stereotaxic Coordinates,* 4th ed. (Paxinos and Watson, 1998), *The Rat Nervous System,* 3rd ed. (Paxinos, 2004), *Chemoarchitectonic Atlas of the Rat Forebrain* (Paxinos *et al.*, 1999) and *Chemoarchitectonic Atlas of the Rat Brainstem* (Paxinos *et al.*, 1999) as the primary guides, supplemented by consideration of the delineations in the Swanson (1992) atlas and specific information on the mouse obtained from previous atlases of the mouse brain, journal articles, and our own observations.

Methods

Surgery All procedures involving live animals were carried out in accordance with accepted ethical principles for animal research and were approved by the relevant Animal Ethics committees of The University of New South Wales and McGill University.

While under sodium pentobarbital anesthesia, the mice were placed in a Kopf small animal stereotaxic instrument. The head was positioned by means of a mouse nose clamp adaptor (Kopf Model 922) supplemented by rat ear bars (Kopf Model 957) placed lightly in the external auditory meatus to locate the interaural line. [NOTE: because the skull of the mouse is extremely fragile and easily compressed, the Kopf mouse holder is designed to be used without ear bars (Slotnick, 1972). If ear bars are used they must be inserted with only a few grams of pressure or else the bones of external auditory meatus or the skull will be crushed. Ear bars can be used successfully if they are inserted so as to merely touch the meatus; any further pressure may lead to breathing difficulties.

The position of the head was adjusted so that the height of the skull surface at bregma and lambda was the same. Lambda was defined as the point of intersection of the best–fit lines passing through the sagittal suture and the left and right portions of the lambdoid suture (see skull diagram below). The mean (±SD) position of bregma was 3.8 (±0.25) mm rostral and 5.8 (±0.48) mm dorsal to the interaural line (IA). Lambda was located 0.41 (±0.26) mm caudal and 5.8 mm dorsal to the interaural line. These values were used to define the scaling of distances on the stereotaxic grid.

To establish the stereotaxic position of brain structures, reference needle tracks were made perpendicular to the horizontal and coronal planes and at predetermined distances from bregma and the interaural line (vertical tracks: 2 mm anterior to bregma and 1 mm anterior to the IA; horizontal track: 3 mm above the IA; all tracks were 1.5 mm lateral to the midline).

The coronal set of the first edition reached only the level of the area postrema. For the second edition we inserted another 7 levels obtained from a different mouse of the same strain and weight.

Histology

After surgery, while still under anesthesia, mice were killed by transcardial perfusion of 20 ml ice–cold phosphate buffered saline (0.1 M PBS, pH 7.3), followed by 20 ml of 4% paraformaldehyde in PBS. The brain was immediately dissected free from the skull and placed dorsal surface down in a small aluminum foil boat containing 1.5% gelatin dissolved in 0.9% saline solution. The aluminum foil boats were made by folding aluminum foil around an appropriately sized rectangular block (20 □ 12 □ 12 mm). The brain was oriented in the boat so that its anterior–posterior axis was parallel to the long axis of the boat. When suspended in the gelatin solution, the brain floats so that its dorsal surface is almost exactly parallel to the flat skull horizontal plane. When the brain was correctly positioned, the boat was placed in a refrigerator for 20 min for the gelatin to jell. The sides of the aluminum boat were then folded down to expose the gelatin block. The foil now formed a tray that was used to lower the gelatin block into dry ice–cooled isopentane. The block was lowered slowly so that freezing proceeded rostrally through the brain, from brainstem to olfactory bulb, and no more than 1–2 mm of unfrozen tissue was below the surface of the isopentane at any time. The frozen block was then stripped of the remaining aluminum foil and attached to the chuck of the microtome with mounting medium. The planar sides of the rectangular block were used to orient the block on the chuck, with the anterior–posterior axis of the brain perpendicular to the surface of the chuck. Only minor adjustment of the orientation of the chuck in the microtome was then necessary to cut sections perpendicular to the horizontal and vertical stereotaxic planes.

The brain was cut at 40–μm thickness sections. Unattached gelatin was brushed away, and the section was taken up into a drop of PBS on the surface of an acid–cleaned and gelatinized slide. Out of every three successive sections, two were taken onto a slide designated for staining for Nissl or AChE. If the first two sections were deemed satisfactory, the third was set aside in ice–cold PBS for parvalbumin immunohistochemistry. The following three sections were designated for staining of AChE or Nissl, respectively, as per the protocol of Paxinos and Watson (1986). A detailed Nissl and AChE staining protocol can be found at the web site http://www.powmri.edu.au.

This technique was developed to reliably obtain precisely oriented sections of relatively undistorted mouse brain in which Nissl– and AChE–stained sections alternated. Fresh or lightly fixed mouse brain tissue is so soft that it distorts under its own weight, but it assumes its normal shape when supported in a viscous fluid. Jelled gelatin solution adequately supports the brain during freezing, but the gelatin causes the sections to curl when they are cut in a freezing microtome. The addition of sodium chloride to the gelatin solution promotes the formation of fine crystals in the gelatin block so that the embedding material shatters when sectioned, and can be gently brushed away from the frozen sections. Freezing must proceed steadily through the long axis of the block. If the whole block is placed directly into the cooling medium, the outside of the block freezes and forms a rigid shell so that when the brain expands during freezing, it

Skull Diagram

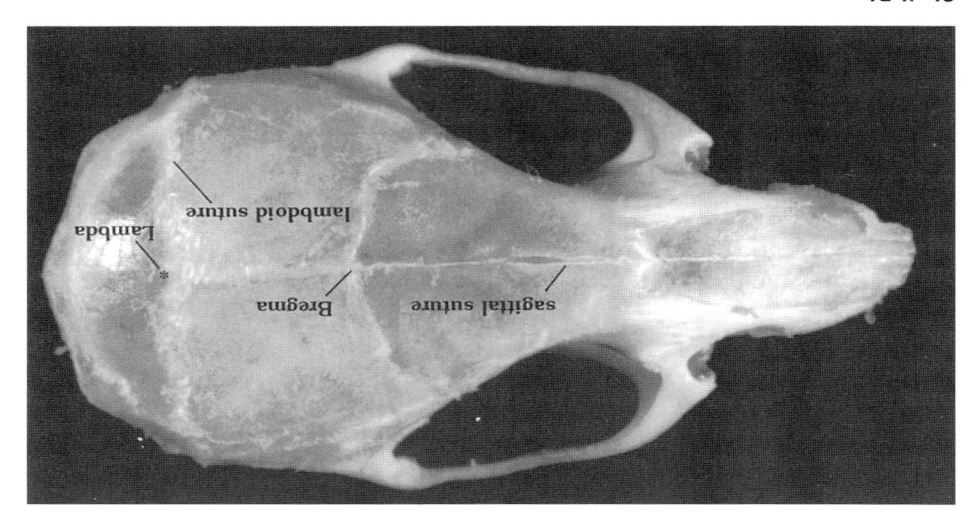

Skull Diagram The dorsal surface of the mouse skull showing the horizontal plane reference points, bregma and lambda. Lambda is defined as the point of intersection of the projection of lines of best fit through the sagittal and lambdoid sutures.

Nomenclature and Abbreviations

The nomenclature and abbreviations used are those that have been employed in *The Rat Brain in Stereotaxic Coordinates, 4th ed.* (Paxinos & Watson, 1998), *The Mouse Brain in Stereotaxic Coordinates* (Franklin & Paxinos, 1997), *Atlas of the Developing Rat Nervous System* (Paxinos et al., 1994), *Chemoarchitectonic Atlas of the Rat Forebrain* (Paxinos et al., 1996b), *Chemoarchitectonic Atlas of the Rat Brainstem* (Paxinos et al., 1999a, b), *The Rhesus Monkey Brain in Stereotaxic Coordinates* (Paxinos et al., 2000), *Atlas of the Human Brain* (Paxinos & Huang, 1995), *Atlas of the Human Brain* (Mai, Asscheuer, & Paxinos, 1997) and *The Chick Brain in Stereotaxic Coordinates* (Puelles et al., in press). We have made a conscious effort to use identical abbreviations for homologous structures in various species so that readers do not have to be burdened with a meaningless task of learning different abbreviations for the same structures. Bowden and Martin (1995) in their atlas of the monkey brain used the same principles for construction of abbreviations as we have used here.

There are more than 20 ways to abbreviate "accumbens nucleus": Acb, ACB, acb, NAS, nas, A, a, Ac, ac, NA, na, AN, an, NAC, nac, ACN, acn, ACU, acu, ACC, Acc. The user of this atlas is welcome to use our abbreviation Acb, the same abbreviation we used in all our atlases.

The principles used in the construction of abbreviations are the same as those used to derive the abbreviations for the elements of the Periodic Table and for the word acetylcholinesterase (AChE).

1. The abbreviations represent the order of words as spoken in English (e.g., DLG = dorsal lateral geniculate nucleus).
2. Capital letters represent nuclei, and lower case letters represent fiber tracts. Thus, the letter "N" has not been used to denote nuclei, and the letter "f" has not been used to denote fiber tracts.
3. The general principle used in the abbreviations of the names of elements in the periodic table was followed: the capital letter representing the first letter of a word in a nucleus is followed by the lower case letter most characteristic of that word (not necessarily the second letter; e.g., Mg = magnesium; Rt = reticular thalamic nucleus).
4. Compound names of nuclei have a capital letter for each part (e.g., LPGi = lateral paragigantocellular nucleus).
5. If a word occurs in a number of the names of structures, it is usually given the same abbreviation (e.g., Rt = reticular thalamus nucleus; RtTg = reticulotegmental nucleus of the pons). Exceptions to this rule are made for well–established abbreviations such as VTA.
6. Abbreviations of brain regions are omitted where the identity of the region in question is clear from its position (CMn = centromedian thalamic nucleus; not CMnTh).
7. Arabic numerals are used instead of Roman numerals in identifying (a) cranial nerves and nuclei (as in the Berman, 1968, atlas), and (b) cerebellar lobules. While the spoken meaning is the same, the detection threshold is lower, ambiguity is reduced, and they are easier to position in small spaces available on diagrams.

Abbreviations representing nuclei and cortical regions are shown in uppercase characters on the left side of the figures, except where structures were only visible on the other side of the section. Fiber tracts and fissures are abbreviated in lowercase characters and are shown on the right side of the figures.

Preparation of Images

Photographs of selected sections were made onto 4 × 5 inch Kodak Plus X or AGFA negatives using a Nikon Multifid photomicroscope, and 20 × 22 inch black and white prints were made of these negatives on AGFA multigrade photographic paper. The delineations and fiducial marks were hand drawn onto acetate tracing paper placed over the photographs. The tracings were scanned, and the resulting images were used as a template for tracing using CorelDraw 5.0 for Windows (coronal set) or Adobe Illustrator 8.0 for Mac OS (sagittal set).

distorts. Sections were taken up into a drop of PBS to eliminate minute bubbles which become trapped under the section if sections are thawed directly onto a slide. With partially fixed mouse brain, the minute bubbles perforate the section when it dries.

Drawings

It was thought that the drawings would be more informative if they were not stylized, and for this reason, with the exception of small adjustments to distorted midlines and cortex, the drawings depict the asymmetries present in the sections. When part of a section was missing or severely distorted, it was drawn in after consideration of sections obtained from other brains.

Coronal Plane

The large number at the bottom left of each figure shows the anteroposterior distance of the corresponding plate from the vertical plane passing through the interaural line. The large number at the bottom right shows the anteroposterior distance of the plate from bregma. The numbers on the left margin show the dorsoventral distance from the horizontal plane passing through the interaural line. The numbers on the right margin show the dorsoventral distance from the horizontal plane passing through bregma and lambda on the surface of the skull. The numbers on the top and bottom margins show the distance of structures from the midline.

Use in Surgery

If it is necessary to insert an electrode into the ventromedial hypothalamic nucleus, Fig. 44 reveals the coordinates with reference to the interaural line to be 2.22 mm anterior to the interaural line, 0.25 mm dorsal to it, and 0.5 mm lateral to the midline. The coordinates of the same structure obtained with reference to bregma are 1.58 mm anterior to bregma, 5.5 mm ventral to it, and 0.5 mm lateral to the midline. The location of bregma and lambda on the mouse skull is shown in the skull diagram.

Use of the David Kopf Stereotaxic Instrument

In our work on this and other stereotaxic atlases we found the David Kopf Small Animal Stereotaxic instrument of high precision. However, no atlas or stereotaxic instrument will compensate for using bregma and lambdoid points inappropriately. These reference skull marks are the midpoints of the curve of best fit along the coronal and the lambdoid suture, respectively. They are not necessarily the points of intersection of these sutures with the midline suture.

The Brain Blocker

Researchers usually wish to block the brain at the same coronal (or sagittal) plane as the atlas plane (flat skull position) so that they can determine most readily the location of their electrode placements and identify structures according to the atlas. The brain blocker depicted in the accompanying photograph allows the blocking of the brain at the stereotaxic plane or the slicing of the brain at 1–mm intervals. (The mouse brain blocker can be purchased by sending an email to brainblocker@hotmail.com; a similarly constructed rat brain blocker is also available.) After blocking, the brain needs to be placed on a chuck with surface parallel to the cryotome knife. For cryotomes without zero–tilt position, this can be achieved by freezing some mounting medium directly onto the chuck and cutting through the mounting medium to create an aligned surface on which to position the blocked tissue.

The Basis of Delineation of Structures

The principal references for our work on the mouse atlas have been *The Rat Brain in Stereotaxic Coordinates*, 4th ed. (Paxinos and Watson, 1998) and *The Rat Nervous System*, 3nd ed. (Paxinos, 2004). This was supplemented by reference to anatomical studies on specific regions in the mouse and consideration of the atlas of Swanson (1992), as well as material on the chemoarchitecture of the rat brain in which each section of the rat brain is depicted after being stained for parvalbumin, calbindin, calbindin, cal-

The Brain Blocker Designed to reproduce the plane of section of *The Mouse Brain in Stereotaxic Coordinates*. To purchase the mouse brain blocker (or a similarly constructed rat brain blocker) email brainblocker@hotmail.com

retinin, Nissl, SMI32, tyrosine hydroxylase, and NADPH–diaphorase (Paxinos et al., 1999a, b).

To resolve ambiguities in delineations based on the Nissl and AChE distribution, we examined mouse brains stained to reveal other neurochemical markers, including substance P, parvalbumin, NADPH–diaphorase, tyrosine hydroxylase, and zinc (Timm's). In many cases, assistance with the construction of boundaries was provided by a number of colleagues who have special knowledge of the anatomy of particular brain regions (see Acknowledgments). We also benefited from our examination of the delineation of structures in a number of rat, mouse, cat, and human brain atlases (Berman and Jones, 1982; Jacobowitz and Palkovits, 1974; König and Klippel, 1981; Olszelwski and Baxter, 1954; Palkovits and Jacobowitz, 1974; Pellegrino et al., 1979; Slotnick and Hersch, 1980; Slotnick and Leonard, 1975; Zilles, 1985; Paxinos et al., 1994; Huang and Paxinos, 1995) and maps of the distribution of neurochemical markers (Arimatsu et al., 1981; Vincent and Kimura, 1992; Rodrigo et al., 1994; Loren et al., 1979; Bartsch, 1991).

Some brief notes on the neuroanatomical literature on the mouse brain and on delineations that differ significantly from the delineations of the rat brain in Paxinos and Watson (1986) are provided below. For a more extensive rationale of the delineations of structures see Paxinos and Watson (1986; 1997).

Forebrain

Olfactory System

Olfactory structures were delineated according to the description of Shipley et al. (1995), and detailed delineations of the mouse olfactory bulb and its connections are found in Bucherelli et al. (1993), Caviness and Sidman (1972), and Shipley and Adamec (1984).

Cortex

Our delineation is based primarily on recent delineations of the cortex of the rat (Zilles and Wree, 1995) and has to be considered as tentative, given that the mouse cortex is different. We supplemented the observations of Zilles and Wree (1995) by examination of the intervening sections from the atlas brain that were immunoreacted for parvalbumin.

For a description of the cytoarchitectonic fields of the mouse cortex, see Caviness (1975). The distribution of choline acetyltransferase and acetylcholinesterase in mouse cortex and other structures is described in Kitt et al. (1994) and Arimatsu et al. (1981). The SMI immunoreactivity of the rat cortex (Paxinos et al., 1999a, b) was critical in suggesting a possible plan for the mouse cortex. Consideration of the delineation of the posterior parietal region in Swanson (1992) was very helpful. That atlas was also assisted by the delineation of the rostral panhandles of the lateral entorhinal cortex and the entorhinal cortex (Swanson, 1992).

Hippocampus and entorhinal cortex delineations were based on Amaral and Witter (1995). The work of Slomianka and Geneser (1991a, b) was helpful because they describe the distribution of AChE in the hippocampal region of the mouse.

Basal Ganglia and Basal Forebrain

The basal ganglia and basal forebrain were delineated by comparison with the rat, guided by the densities of Nissl, AChE staining, and substance P immunoreactivity. Refer to Heimer et al. (1995) for a discussion of the organization description of the basal ganglia in the rat. The cholinergic innervation of the basal forebrain in the mouse has been described by Kitt et al. (1994).

Interstitial Nucleus of the Posterior Limb of the Anterior Commissure (IPAC)

Dorsal to the anterior amygdaloid area and ventral to the fundus striati lies a region that is less cellular and less reactive for AChE than the fundus striati (Figs. 22–24 in Paxinos and Watson, 1986). In the rat, this region is shown to be reactive for substance P (Fig. 3E of Haber and Nauta, 1983), and this reactivity is continuous with that of the globus pallidus. The region has similar characteristics in the mouse. Paxinos and Watson (1986) named this region the substriatal area (SStr). More recently it has been suggested (Alheid et al., 1995) that this region should be named the interstitial nucleus of the posterior limb of the anterior commissure (IPAC) on the basis of its association with the posterior limb of the anterior commissure throughout most of its course, and we have followed this proposal (Figs. 27–36). This region is distinctive in receiving particular afferents from the amygdala. It is densely stained by Timm's reaction and AChE, and the dorsal and lateral part of the region is densely supplied with substance P terminals. In the rat the IPAC is distinguished by receptor binding for calcitonin (Skofitsch and Jacobowitz, 1992) and vasopressin (Tribollet, 1992).

Substantia Innominata (SI)
Following the carving out of the ventral pallidum from the substantia innominata, the remainder (posterior part) was called the sublenticular part of the SI (Heimer et al., 1985). Paxinos and Watson (1986) noticed that a strip of substantia innominata persists rostrally bounded by the ventral pallidum, the lateral preoptic area, the nucleus of the horizontal limb of the diagonal band, and the magnocellular preoptic nucleus. They retained the term substantia innominata for this panhandle. It has been argued that the cells of this region, traditionally called substantia innominata, represent the rostral end of a continuous column of cells stretching from the amygdala in the temporal lobe to the forebrain (Alheid et al., 1995). Alheid et al. (1995) have named this region the sublenticular extended amygdala and divided it into two parts: a medial division (SLEAM) and a central division (SLEAC). In the mouse we

have retained the more traditional delineation of the substantia innominata, but also show the proposed division into medial and central sublenticular extended amygdala (Figs. 34–39). The central division corresponds to the anterodorsal substantia innominata (De Olmos et al., 1985), or what Grove (1988a, b) termed the dorsal subdivision of the SI, which is lightly stained with AChE and Timm's reaction. The medial division corresponds to Grove's ventral division of the substantia innominata, or the posteroventral substantia innominata of De Olmos et al. (1985), and it stains moderately for substance P, AChE, and Timm's reaction.

Basal Nucleus (B)   The cholinergic cells of the basal nucleus of Meynert straddle the border between the globus pallidus and the internal capsule, invading both of these structures (Figs. 34–36). For a general description of the distribution of cholinergic neurons in the basal forebrain in the mouse, see Kitt et al. (1994).

Subthalamic Nucleus   The subthalamic nucleus (Figs. 45–52) is prominent in the mouse. Laterally it is lens shaped, as in the rat, but medially it becomes more rounded than in the rat. At the caudal limit of the subthalamic nucleus there is a tightly packed group of cells that extends caudally from the posterior limit of the magnocellular nucleus of the lateral hypothalamus to about 200 mm rostral to the substantia nigra (Sturrock, 1991; Broadwell and Bleier, 1977). The appearance of this distinct cluster of cells is quite different from the subthalamic nucleus in the rat, monkey (Paxinos et al., 1996), or human (Paxinos and Huang, 1995) and we have not assimilated it to the subthalamic nucleus as have some authors (Sturrock, 1991; Broadwell and Bleier, 1977). Wang (1995) has suggested that the cell group should be distinguished from the subthalamic nucleus as the parasubthalamic nucleus (PSTh), and we have followed his suggestion.

Hypothalamus

Our delineation of the hypothalamus is based on the description of Simerly (1995). The general cytoarchitecture of the mouse hypothalamus has been described by Broadwell and Bleier (1977). The distribution of aromatase–immunoreactive cells in the hypothalamus and other areas has been mapped by Foidart et al. (1995). See Ruggiero et al. (1984) and Baker et al. (1993) for a description of catecho–laminergic cells in the hypothalamus of the mouse.

The medial preoptic area was delineated in accordance with Simerly (1995) except for the ventromedial preoptic nucleus (VMPO), which was delineated in accordance with Elmquist et al. (1996). See Paxinos and Watson (1997) for additional discussion of this issue.

Anteroventral Periventricular Nucleus (AVPe)   This small compact, rostral cell group (Figs. 26–30) was named the medial preoptic nucleus by Bleier et al. (1979), but most other researchers reserve this term for the larger nucleus that succeeds it posteriorly. Both of these nuclei are part of the sexually dimorphic region (Bleier et al., 1982). For the mouse brain we have adopted the term "anteroventral periventricular nucleus" (Swanson, 1992), which is descriptive of its position and has become more widely used.

Anterodorsal Preoptic Nucleus (ADP)   This nucleus is delineated on the basis that it is distinctly labeled by tyrosine hydroxylase immunoreactivity in the mouse (Ruggiero et al., 1984). It lies just ventral to decussation of the anterior commissure, lateral to the periventricular nucleus and median preoptic nucleus, and medial to the ventral medial division of the bed nucleus of the stria terminalis (Figs. 30 and 31). In Nissl–stained sections it contains small, round to ovoid, lightly staining cells. In sections treated to reveal tyrosine hydroxylase–like immunore–activity, the nucleus appears as a tongue protruding laterally from the third ventricle, dorsal to the medial preoptic nucleus. In the rat and other mammalian orders, the region does not contain catecholaminergic cells (see Ruggiero et al., 1984). We have followed the suggestion of Ruggiero et al. (1984) and identified the catecholamine cell group as an extension of A14.

Anterior Hypothalamic Nucleus (AH)   Contro–versy exists over the location of the anterior hypothalamic nucleus. We followed Paxinos and Watson (1986) in dividing the AH into three parts that merge into each other. However, our most anterior part is immediately dorsal to that which was identified by Saper et al. (1978) as the "anterior part of the anterior hypothalamic nucleus." The "anterior part of the anterior hypothalamic nucleus" of Saper et al. (1978) corresponds to the "lateral anterior hypothalamic nucleus" of Bleier et al. (1979). Our "anterior part of the anterior hypothalamic nucleus" is immediately dorsal to the "lateral anterior hypothalamic nucleus" and ventral to the bed nucleus of the stria terminalis (Figs. 34–36). The central part of the anterior hypothalamic nucleus of Saper et al. (1978) and our Figs. 37 and 38 correspond to the dorsal tuberal nucleus of Bleier et al. (1979). The posterior part of the anterior hypothalamic nucleus is immediately ventral to the most caudal part of the paraventricular hypothalamic nucleus (Figs. 39–42). Within the dorsal aspects of the posterior part of the AH there is a small, dense cluster of cells that is positive for glutamic acid decarboxylase in the rat. Paxinos and Watson (1986) named it the stigmoid (dot–like) nucleus (Stg; Fig. 41).

Magnocellular Lateral Hypothalamic Nucleus (MCLH)   A magnocellular nucleus with AChE–positive cells is found ventromedial to the internal capsule at the antero-posterior level of the perifornical and subincertal nuclei (Figs. 43–48). This nucleus has been termed the lateral tuberal nucleus by Bleier et al. (1979), but this term is problematic given that it refers to another structure in primate literature. We have followed Paxinos and Watson (1986) and used the name magnocellular lateral hypothalamic nucleus.

Supraoptic and Paraventricular Nuclei   Delineations are according to Armstrong

(1985). See Silverman and Pickard (1983) for a description of these nuclei in the mouse (Figs. 36–41).

Thalamus

The delineatation of the thalamic nuclei is based on its characteristics in the rat as described by Price (1995) and Paxinos and Watson (1986). For descriptions of the thalamus of the mouse, see Baker *et al.* (1993) and Carvell and Simons (1987). The nomenclature follows Price (1995). The gelatinosus nucleus (G in Paxinos and Watson) is designated as the submedial nucleus (O'Gorman and Sidman, 1996) (Sub; Figs. 39–46) because this term has become more widely used. The small–celled region in the medial part of the ventroposteromedial nucleus (designated "Gu" in Paxinos and Watson, 1986) is recognized as having additional functions to gustation, but we label it "gustatory" for its principal function (Figs. 47–50).

We follow Paxinos and Watson (1986) in delineating the ethmoid nucleus (Eth; Figs. 52–54) and retroethmoid nucleus (Reth; Figs. 53–55).

Hindbrain

Periacqueductal Gray The periaqueductal gray (Figs. 52–74) is divided into longitudinal columns as proposed by Andrezik and Beitz (1985) and Carrive *et al.* (1987) (see also Carrive, 1993). Delineation was assisted by the fact that the dorsolateral lateral column shows a darker acetylcholinesterase reaction than the other columns, whereas the dorsomedial PAG is least reactive (Graybiel and Illing, 1994). In NADPH–diaphorase–stained sections the dorsolateral column also shows more intense reactivity (Carrive and Paxinos, 1994). The NADPH–diaphorase reactive cells and plexus of reactive fibers and terminals in the supraoculomotor cap (SU3C; Figs. 61–66) (Carrive and Paxinos, 1994) are clearly visible in the mouse (unpublished observations).

Dorsal Raphe The subdivision of the dorsal raphe was guided by recent descriptions of the raphe in the human (Huang and Paxinos, 1995).

Parabrachial Nucleu The subdivisions of the parabrachial nucleus (Figs. 73–78) were delineated following the description in the rat by Saper (1995).

Nucleus of the Solitary Tract The delineation of the solitary tract nucleus (Figs. 83–93) was based on Luan Ling Zhang (1999) and on McRitchie (1992) and is similar to the delineations in the rat described by Norgren (1995) for the rostral, gustatory part and by Saper (1995) for the caudal, autonomic part.

Superior Olive The superior olive delineation was based on Ruigrok and Cella (1995). A similar delineation in the normal and "lurcher" mutant mouse can be found in Heckroth and Eisenman (1991).

Cerebellum and Cerebellar Nuclei The cerebellar nuclei are delineated according to Voogd (1995). It should be noted that the folial pattern of the cerebellum is different in different strains of mice and that the C57BL/J6 strain has one of the simplest patterns (Inouye and Oda, 1980).

References

Alheid, G. F., De Olmos, J. S., and Beltramino, C. A. (1995). Amygdala and extended amygdala. In G. Paxinos (Ed.), *The Rat Nervous System*, 2nd ed. (pp. 495–578). Academic Press, San Diego.

Amaral, D. G., and Witter, M. P. (1995). Hippocampal formation. In G. Paxinos (Ed.), *The Rat Nervous System*, 2nd ed. (pp. 443–493). Academic Press, San Diego.

Andrezik, J. A., and Beitz, A. J. (1985). Reticular formation, central gray and related tegmental nuclei. In G. Paxinos (Ed.), *The Rat Nervous System. Vol. 2. Hindbrain and Spinal cord* (pp. 1–28). Academic Press, Sydney.

Arimatsu, Y., Seto, A., and Amano, T. (1981). An atlas of alphabungarotoxin binding sites and structures containing acetylcholinesterase in the mouse central nervous system. *J. Comp. Neurol.* **198**, 603–631.

Armstrong, W. E. (1995). Hypothalamic supraoptic and paraventricular nuclei. In G. Paxinos (Ed.), *The Rat Nervous System*, 2nd ed. (pp. 377–390). Academic Press, San Diego.

Baker, H., Joh, T.H., Ruggiero, D. A., and Reis, D. J. (1993). Variations in number of dopamine neurons and tyrosine hydroxylase activity in hypothalamus of two mouse strains. *J. Neurosci.* **3**, 832–843.

Bartsch, D., and Mai, J. K. (1991). Distribution of the 3-fucosyl-N-acetyl-lactosamine (FAL) epitope in the adult mouse brain. *Cell Tissue Res.* **263**, 353–366.

Berman, A.L., and Jones, E.G. (1982). *The Thalamus and Basal Telencephalon of the Cat.* University of Wisconsin Press, Madison, WI.

Bleier, R., Byne, W., and Siggelkow, I. (1982). Cyto-architectonic sexual dimorphisms of the medial preoptic and anterior hypothalamic areas in guinea pig, rat, hamster, and mouse. *J. Comp. Neurol.* **212**, 118–130.

Bleier, R., Cohn, P., and Siggelkow, I. R. (1979). A cytoarchitectonic atlas of the hypothalamus and hypothalamic third venytricle of the rat. In P. J. Morgane and J. Panksepp (Eds.), *Handbook of the Hypothalamus. Vol. 1. Anatomy of the Hypothalamus* (pp. 137–220).

Broadwell, R. D., and Bleier, R. (1977). A cytoarchitectonic atlas of the mouse hypothalamus. *J. Comp. Neurol.* **167**, 315–340.

Carrive, P. (1993). The periaqueductal gray and defensive behavior: Functional representation and neuronal organization. *Behav. Brain Res.* **58**, 27–47.

Carrive, P., Dampney, R. A. L., and Bandler, R. (1987). Excitation of neurones in a restricted portion of the midbrain periaqueductal gray elicits both behavioural and cardiovascular components of the defence reaction in the unanaesthetised decerebrate cat. *Neurosci Lett*, **81**, 273–278.

Carrive, P., and Paxinos, G. (1994). The supraoculomotor cap: a region revealed by NADPH diaphorase histochemistry. *NeuroReport* **5**, 2257–2260.

Carvell, G. E., and Simons, D. J. (1987). Thalamic and corticocortical connections of the second somatic sensory area of the mouse. *J. Comp. Neurol.* **265**, 409–427.

Caviness, V. S. (1975). Architectonic map of neocortex of the normal mouse. *J. Comp. Neurol.* **164**, 247–264.

Caviness, V. S., and Sidman, R. L. (1972). Olfactory structures of the forebrain in the reeler mutant mouse. *J. Comp. Neurol.* **145**, 85–104.

De Olmos, J., Alheid, G. F., and Beltramino, C. A. (1985). Amygdala. In G. Paxinos (Ed.), *The Rat Nervous System. Vol. 1. Forebrain and Midbrain* (pp. 223–334). Academic Press, Sydney.

Elmquist, J. K., Scammell, T. E., Jacobson, C. D., and Safer, C. B. (1996). Distribution of fos-like immunoreactivity in the rat brain following lipopolysaccharide administration. *J. Comp. Neurol.* **371**, 85–103.

Foidart, A., Harada, N., and Balthazart, J. (1995). Aromatase immunoreactive cells are present in mouse brain areas that are known to express high levels of aromatase activity. *Cell Tissue Res.* **280**, 561–574.

Graybiel, A. M., and Illing, R. B. (1994). Enkephalinpositive and acetylcholinesterasepositive patch systems in the superior colliculus have matching distributions but distinct developmental histories. *J. Comp. Neurol.* **340**, 297–310.

Grove, E. A. (1988a). Efferent connections of the substantia innominata in the rat. *J. Comp. Neurol.* **277**, 347364.

Grove, E. A. (1988b). Neural associations of the substantia innominata in the rat: Afferent connections. *J. Comp. Neurol.* **277**, 315–346.

Heckroth, J. A., and Eisenman, L. M. (1991). Olivary morphology and olivocerebellar topography in adult lurcher mutant mice. *J. Comp. Neurol.* **312**, 641–651.

Heimer, L., Alheid, G. F., and Zaborsky, L. (1985). Basal ganglia. In G. Paxinos (Ed.), *The Rat Nervous System* (pp. 37–86). Academic Press, Sydney.

Heimer, L., Zahm, D. S., and Alheid, G. F. (1995). Basal ganglia. In G. Paxinos (Ed.), *The Rat Nervous System*, 2nd ed. (pp. 579–628). Academic Press, San Diego.

Huang, S., and Paxinos, G. (1995). *Atlas of the Human Brainstem*. Academic Press, San Diego.

Inouye, M., and Oda, S. I. (1980). Strain-specific variations in the folial pattern of the mouse cerebellum. *J. Comp. Neurol.* **190**, 357–362.

Jacobowitz, D. M., and Palkovits, M. (1974). Topographic atlas of catecholamine and acetylcholinesterase-containing neurons in the rat brain. I. Forebrain (telencephalon, diencephalon). *J. Comp. Neurol.* **157**, 13–28.

Kitt, C. A., Höhmann, C., Coyle, J. T., and Price, D. (1994). Cholinergic innervation of mouse forebrain structures. *J. Comp. Neurol.* **341**, 117–129.

König, J. F. R., and Klippel, R. A. (1963). *The Rat Brain: A Stereotaxic Atlas of the Forebrain and Lower Parts of the Brain Stem*. Williams & Wilkins, Baltimore.

Loren, I., Emson, P.C., Fahrenkrug, J., Björklund, A., Alumets, J., Hakanson, and Sundler, F. (1979). Distribution of vasoactive intestinal polypeptide in the rat and mouse brain. *Neuroscience* **4**, 1953–1976.

Mai, J. K., Assheuer, J., and Paxinos, G. (2003). *Atlas of the Human Brain*, 2nd ed., Academic Press, San Diego.

McRitchie, D. A. (1992). *Cytoarchitecture and Chemical Neuro–anatomy of the Nucleus of the Solitary tract: Comparative and Experimental Studies in the Human and the Rat*. Doctoral Dissertation, School of Anatomy, University of New South Wales.

Meesen, H., and Olszewski, J. (1949). *A Cytoarchitectonic Atlas of the Rhombecephalon of the Rabbit*. Karger, Basel.

Mitro, A., and Palkovits, M. (1981). The morphology of the rat brain ventricles, ependyma and periventricular structures. *Biblthca Anat.* **21**, 1–110.

Montemurro, D. G., and Dukelow, R. H. (1972). *A Stereotaxic Atlas of the Diencephalon and Related Structures of the Mouse*. Futura, Mount Kisco, NY.

Norgren, R. (1995). Gustatory system. In G. Paxinos (Ed.), *The Rat Nervous System*, 2nd ed. (pp. 751–771). Academic Press, San Diego.

O'Gorman, S., and Sidman, R. L. (1996). Degeneration of thalamic neurons in "Purkinje cell degeneration" mutant mice. I. Distribution of neuron loss. *J. Comp. Neurol.* **234**, 277–297.

Olszelwski, J., and Baxter, D. (1954). *Cytoarchitecture of the Human Brain Stem*. Karger, Basel.

Palkovits, M., and Jacobowitz, D. M. (1974). Topographic atlas of the catecholamine and acetyl-cholinesterasecontaining neurons in the rat brain. II. Hindbrain (mesencephalon, rhombencephalon). *J. Comp. Neurol.* **157**, 29–42.

Paxinos, G (Ed) (2004), *The Rat Nervous System*, 2nd ed., Academic Press, San Diego.

Paxinos, G., and Watson, C. (1986). *The Rat Brain in Stereotaxic Coordinates*, 2nd ed. Academic Press, Sydney.

Paxinos, G., and Huang, X. F. (1995). *Atlas of the Human Brainstem*, Academic Press, San Diego.

Paxinos, G., and Watson, C. (1997). *The Rat Brain in Stereotaxic Coordinates*, 3rd ed. Academic Press, San Diego.

Paxinos, G., and Watson, C. (1998). *The Rat Brain in Stereotaxic Coordinates*, 4th ed. Academic Press, San Diego.

Paxinos, G., Ashwell, K. W. S., and Törk, I. (1994). *Atlas of the Developing Rat Nervous System*. Academic Press, San Diego.

Paxinos, G., Carrive, P., Wang, H. Q., and Wang, P. Y. (1999). *Chemoarchitecture of the Rat Brainstem*. Academic Press, San Diego.

Paxinos, G., Huang, X.-F., and Toga, A. W. (2000). *The Rhesus Monkey Brain in Stereotaxic Coordinates*. Academic Press, San Diego.

Paxinos, G., Kus, L., Ashwell, K. W., and Watson, C. (1999). *Chemoarchitecture of the Rat Forebrain*. Academic Press, San Diego.

Pellegrino, L. J., Pellegrino, A. S., and Cushman, A. J. (1979). *A Stereotaxic Atlas of the Rat Brain*. Plenum, New York.

Price, J. L. (1995). Thalamus. In G. Paxinos (Ed.), *The Rat Nervous System*, 2nd ed. (pp. 629–648). Academic Press, San Diego.

Puelles, L., Martinez-de-la-Torre, Marinez, S., Watson, C., and Paxinos G. (in press). The Chick Brain in Stereotaxic Coordinates, Academic Press, San Diego.

Rodrigo, J., Springall, D. R., Uttenthal, O., Bentura, M. L., Abadia-Molina, F., Riveros-Moreno, V., Martinez-Murillo, R., Polak, J. M., and Moncada, S. (1994). Localization of nitric oxide synthase in the adult rat brain. *Phil. Trans. R. Soc. Lond. B* **345**, 175–221.

Ruggiero, D. A., Baker, H., Joh, T. H., and Reis, D. J. (1984). Distribution of catecholamine neurons in the hypothalamus and preoptic region of mouse. *J. Comp. Neurol.* **223**, 556–582.

Ruigrok, T. J. H., and Cella, F. (1995). Precerebellar nuclei and red nucleus. In G. Paxinos (Ed.), *The Rat Nervous System*, 2nd ed. (pp. 277–208). Academic Press, San Diego.

Saper, C. B. (1995). Central autonomic system. In G. Paxinos (Ed.), *The Rat Nervous System*, 2nd ed. (pp. 107–135). Academic Press, San Diego.

Saper, C. B., Swanson, L. W., and Cowan, W. M. (1978). The efferent connections of the anterior hypothalamic area of the rat, cat and monkey. *J. Comp. Neurol.* **182**, 575–600.

Shipley, M. T., and Adamec, G. D. (1984). The connections of the mouse olfactory bulb: A study using orthograde and retrograde transport of wheat germ agglutinin conjugated horseradish peroxidase. *Brain Res. Bull.* **12**, 669–688.

Shipley, M. T., McLean, J. H., and Ennis, M. (1995). Olfactory system. In G. Paxinos (Ed.), *The Rat Nervous System*, 2nd ed. (pp. 899–926). Academic Press, San Diego.

Silverman, A., and Pickard, G. E. (1983). The hypothalamus. In P. C. Emson (Ed.), *Chemical Neuroanatomy* (pp. 295–336). Raven Press, New York.

Simerly, R. B. (1995). Anatomical substrates of hypothalamic integration. In G. Paxinos (Ed.), *The Rat Nervous System*, 2nd ed. (pp. 353–376). Academic Press, San Diego.

Skofitsch, G., and Jacobowitz, D. M. (1992). Calcitonin and calcitonin generelated peptide: Receptor binding sites in the central nervous system. In A. Björklund, T. Hökfelt, and M. J. Kubar (Eds.), *Handbook of Chemical Neuroanatomy: Neuropeptide Receptors in the CNS* (pp. 97–145). Elsevier, Amsterdam.

Slomianka, L., and Geneser, F. A. (1991a). Distribution of acetylcholinesterase in the hippocampal region of the mouse. I. Entorhinal area, parasubiculum, retrosplenial area, and presubiculum. *J. Comp. Neurol.* **303**, 339–354.

Slomianka, L., and Geneser, F. A. (1991b). Distribution of acetylcholinesterase in the hippocampal region of the mouse. II. Subiculum and hippocampus. *J. Comp. Neurol.* **312**, 525–536.

Slotnick, B. M. (1972). Stereotaxic surgical techniques for the mouse. *Physiol. Behav.* **8**, 139–142.

Slotnick, B. M., and Hersch, S. (1980). A stereotaxic atlas of the rat olfactory system. *Brain Res. Bull.* **5**, 1–55.

Slotnick, B. M., and Leonard, C. M. (1975). *A Stereotaxic Atlas of the Albino Mouse Forebrain.* U.S. Department of Health, Education and Welfare, Rockville, MD.

Sturrock, R. R. (1991). Stability of neurons in the subthalamic and entopeduncular nuclei in the ageing mouse brain. *J. Anat.* **179**, 67–73.

Swanson, L. W. (1992). *Brain Maps: Structure of the Rat Brain.* Elsevier, Amsterdam.

Tribollet, E. (1992). Vasopressin and oxytocin receptors in the rat brain. In A. Björklund, T. Hökfelt, and M. J. Kubar (Eds.), *Handbook of Chemical Neuroanatomy: Neuropeptide Receptors* in the CNS (pp. 289–320). Elsevier, Amsterdam.

Vincent, S. R., and Kimura, H. (1992). Histochemical mapping of nitric oxide synthase in the rat brain. *Neuroscience* **46**, 755–784.

Voogd, J. (1995). Cerebellum. In G. Paxinos (Ed.), *The Rat Nervous System*, 2nd ed. (pp. 309–350). Academic Press, San Diego.

Wahlsten, D., Hudspeth, W. J., and Bernhardt, K. (1975). Implications of genetic variation in mouse brain structure for electrode placement by streotaxic surgery. *J. Comp. Neurol.* **162**, 519–531.

Wang, P. Y. (1995). *Outlines and Atlas of Learning Rat Brain.* Northwest University Xi'an, Xi'an, China.

Zhang, L. L. (1999). *Development of the rat nucleus of the solitary tract and its visceral afferents.* Unpublished Ph.D thesis, The University of New South Wales.

Zilles, K. (1985). *The Cortex of the Rat: A Stereotaxic Atlas.* SpringerVerlag, Berlin.

Zilles, K., and Wree, A. (1995). Cortex: Areal and lamina structure. In G. Paxinos (Ed.), *The Rat Nervous System*, 2nd ed. (pp. 649–685). Academic Press, San Diego.

List of Structures

Names of the structures are listed in alphabetical order. Each name is followed by abbreviation of the structure.

1st Cerebellar lobule   1Cb
3rd ventricle   3V
4&5th Cerebellar lobules   4&5Cb
4th ventricle   4V
5th Cerebellar lobule   5Cb
6th Cerebellar lobule   6Cb
7th Cerebellar lobule   7Cb
8th Cerebellar lobule   8Cb
9th Cerebellar lobule   9Cb
10th Cerebellar lobule   10Cb

A

A11 dopamine cells   A11
A12 dopamine cells   A12
A13 dopamine cells   A13
A14 dopamine cells   A14
accessory abducens/facial nucleus   Acs6/7
accessory facial nucleus   Acs7
accessory neurosecretory nuclei   Acc
accessory nucleus of the ventral horn   Acs
accessory olfactory bulb   AOB
accessory olfactory tract   aot
accessory optic tract   aopt
accessory trigeminal nucleus   Acs5
accumbens nucleus   Acb
accumbens nucleus, core   AcbC
accumbens nucleus, shell   AcbSh
acoustic radiation   ar
agranular insular cortex, dorsal part   AID
agranular insular cortex, posterior part   AIP
agranular insular cortex, ventral part   AIV
alveus of the hippocampus   alv
ambiguus nucleus   Amb
amygdalohippocampal area, anterolateral part   AHiAL
amygdalohippocampal area, posterolateral   AHiPL
amygdalohippocampal area, posteromedial part   AHiPM
amygdaloid fissure   af
amygdalopiriform transition area   APir
amygdalostriatal transition area   AStr
angular thalamic nucleus   Ang
ansa lenticularis   al
ansoparamedian fissure   apmf
anterior amygdaloid area   AA
anterior amygdaloid area, dorsal part   AAD
anterior amygdaloid area, ventral part   AAV
anterior commissural nucleus   AC
anterior commissure   ac
anterior commissure, intrabulbar part   aci
anterior cortical amygdaloid nucleus   ACo
anterior hypothalamic area   AH
anterior hypothalamic area, anterior part   AHA
anterior hypothalamic area, central part   AHC
anterior hypothalamic area, posterior part   AHP
anterior interposed nucleus   AInt
anterior lobe cerebellum   Ant
anterior olfactory nucleus, dorsal part   AOD
anterior olfactory nucleus, external part   AOE
anterior olfactory nucleus, lateral part   AOL
anterior olfactory nucleus, medial part   AOM
anterior olfactory nucleus, posterior part   AOP
anterior periformical nucleus   APF
anterior preoptic nucleus   APO
anterior pretectal nucleus   APT
anterior pretectal nucleus, dorsal part   APTD
anterodorsal preoptic nucleus   ADP
anterodorsal thalamic nucleus   AD
anteromedial thalamic nucleus   AM
anteromedial thalamic nucleus, ventral part   AMV
anteroventral preoptic nucleus   AVPO
anteroventral thalamic nucleus   AV
anteroventral thalamic nucleus, dorsomedial part   AVDM
anteroventral thalamic nucleus, ventrolateral part   AVVL
arcuate hypothalamic nucleus   Arc
arcuate hypothalamic nucleus, dorsal part   ArcD
arcuate hypothalamic nucleus, lateral part   ArcL
arcuate hypothalamic nucleus, medial part   ArcM
arcuate hypothalamic nucleus, medial posterior part   ArcMP
area postrema   AP
artery   a
ascending fibers of the facial nerve   asc7

B

B4 serotonin cells   B4
B9 serotonin cells   B9
Barrington's nucleus   Bar
basal nucleus (Meynert)   B
basilar artery   bas
basolateral amygdaloid nucleus   BL
basolateral amygdaloid nucleus, anterior part   BLA
basolateral amygdaloid nucleus, posterior part   BLP
basomedial amygdaloid nucleus   BM
basomedial amygdaloid nucleus, anterior part   BMA
basomedial amygdaloid nucleus, posterior part   BMP
bed nucleus of stria terminalis, supracapsular part   BSTS
bed nucleus of the accessory olfactory tract   BAOT
bed nucleus of the anterior commissure   BAC
bed nucleus of the stria terminalis, intermediate division   BSTI
bed nucleus of the stria terminalis, intraamygdaloid division   BSTIA
bed nucleus of the stria terminalis, lateral division   BSTL
bed nucleus of the stria terminalis, lateral division, dorsal part   BSTLD
bed nucleus of the stria terminalis, lateral division, juxtacapsular part   BSTLJ
bed nucleus of the stria terminalis, lateral division, posterior part   BSTLP
bed nucleus of the stria terminalis, lateral division, ventral part   BSTLV
bed nucleus of the stria terminalis, medial division   BSTM
bed nucleus of the stria terminalis, medial division, posterior part   BSTMP
bed nucleus of the stria terminalis, medial division, posterolateral part   BSTMPL
bed nucleus of the stria terminalis, medial division, posteromedial part   BSTMPM
bed nucleus of the stria terminalis, medial division, ventral part   BSTMV
bed nucleus of the stria terminalis, ventral division   BSTV
blood vessel   bv
Botzinger complex   Bo
brachium of the inferior colliculus   bic
brachium of the superior colliculus   bsc
brachium pontis (stem of middle cerebellar peduncle)   bp

C

C2 adrenaline cells   C2
C3 adrenaline cells   C3
caudal interstitial nucleus of the medial longitudinal fasciculus   CI
caudal linear nucleus of the raphe   CLi
caudal periolivary nucleus   CPO
caudal tuberomammillary nucleus   CTM
caudate putamen (striatum)   CPu
caudoventral respiratory group   CVRG
caudoventrolateral reticular nucleus   CVL
cell bridges of the ventral striatum   CB
central amygdaloid nucleus, medial division, anterodorsal part   CeMAD
central amygdaloid nucleus, medial division, anteroventral part   CeMAV
central amygdaloid nucleus   Ce
central amygdaloid nucleus, capsular part   CeC
central amygdaloid nucleus, medial division   CeM
central amygdaloid nucleus, medial posteroventral part   CeMPV
central canal   CC
central gray of the pons   CGPn
central gray, alpha part   CGA
central gray, beta part   CGB
central medial thalamic nucleus   CM
central tegmental tract   ctg
centrolateral thalamic nucleus   CL
cerebellar commissure   cbc
cerebellar lobule   6a
cerebellar lobule   6b
cerebellar lobules   4&5
cerebellum   Cb
cerebral cortex   Cx
cerebral peduncle, basal part   cp
choroid plexus   chp
cingulate cortex, area 1   Cg1
cingulate cortex, area 2   Cg2
circular nucleus   Cir
claustrum   Cl
cochlear root of the vestibulocochlear nerve   8cn
commissural stria terminalis   cst
commissure of the inferior colliculus   cic
commissure of the lateral lemniscus   cll
commissure of the superior colliculus   csc
copula of the pyramis   Cop
corpus callosum   cc
cortex–amygdala transition zone   CxA
crus 1 of the ansiform lobule   Crus1
cuneate fasciculus   cu
cuneate nucleus   Cu
cuneiform nucleus   CnF

xix

D

dcs   decussation of the superior cerebral peduncle
dcw   deep cerebral white matter
DPG   deep gray layer of the superior colliculus
DPMe   deep mesencephalic nucleus
DpWh   deep white layer of the superior colliculus
DG   dentate gyrus
D3V   dorsal 3rd ventricle
DC   dorsal cochlear nucleus
DEn   dorsal endopiriform nucleus
df   dorsal fornix
dhc   dorsal hippocampal commissure
DA   dorsal hypothalamic area
Do   dorsal hypothalamic nucleus
DLG   dorsal lateral geniculate nucleus
dlo   dorsal lateral olfactory tract
dlf   dorsal longitudinal fasciculus
10N   dorsal motor nucleus of vagus
DPGi   dorsal paragigantocellular nucleus
DP   dorsal peduncular cortex
DPPn   dorsal peduncular pontine nucleus
DPO   dorsal periolivary region
DR   dorsal raphe nucleus
DRC   dorsal raphe nucleus, caudal part
DRD   dorsal raphe nucleus, dorsal part
DRI   dorsal raphe nucleus, interfascicular part
DRV   dorsal raphe nucleus, ventral part
dsc   dorsal spinocerebellar tract
DTg   dorsal tegmental nucleus
DTgC   dorsal tegmental nucleus, central part
DTgP   dorsal tegmental nucleus, pericentral part
DTT   dorsal transition zone
DTM   dorsal tuberomammillary nucleus
DLO   dorsolateral orbital cortex
DLPAG dorsolateral periaqueductal gray
DLPn   dorsolateral pontine nucleus
DMC   dorsomedial hypothalamic nucleus, compact part
DMD   dorsomedial hypothalamic nucleus, dorsal part
DMV   dorsomedial hypothalamic nucleus, ventral part
DMPn   dorsomedial pontine nucleus
DMSp5 dorsomedial spinal trigeminal nucleus
DMSp5V dorsomedial spinal trigeminal nucleus, ventral part
DI   dysgranular insular cortex

E

Ect   ectorhinal cortex
E5   ectotrigeminal nucleus
EW   Edinger–Westphal nucleus
Ent   entorhinal cortex
E/OV   ependymal and subendymal layer/olfactory ventricle
EF   epifascicular nucleus
ELm   epilemniscal nucleus
EMi   epimicrocellular nucleus
ERS   epinthrospinal nucleus
Eth   ethmoid thalamic nucleus
ECu   external cuneate nucleus
eml   external medullary lamina
EPl   external plexiform layer of the olfactory bulb
exc   extreme capsule

F

FVe   F cell group of the vestibular complex
7n   facial nerve or its root
7N   facial nucleus
7DL   facial nucleus, lateral subnucleus
7DM   facial nucleus, dorsomedial subnucleus
7DL   facial nucleus, dorsolateral subnucleus
7L   facial nucleus, lateral subnucleus
7VI   facial nucleus, ventral intermediate subnucleus
7VM   facial nucleus, ventromedial subnucleus
fr   fasciculus retroflexus
FC   fasciola cinereum
CA1   field CA1 of hippocampus
CA2   field CA2 of hippocampus
CA3   field CA3 of hippocampus
fi   fimbria of the hippocampus
Fl   flocculus
fmj   forceps major of the corpus callosum
fmi   forceps minor of the corpus callosum
FL   forelimb area, cortex
f   fornix
FrA   frontal association cortex

G

Ge5   gelatinous layer of the caudal spinal trigeminal nucleus
Gem   gemini hypothalamic nucleus
g7   genu of the facial nerve
Gi   gigantocellular reticular nucleus
GiA   gigantocellular reticular nucleus, alpha part
GiV   gigantocellular reticular nucleus, ventral part
GlA   glomerular layer of the accessory olfactory bulb
9n   glossopharyngeal nerve
Gr   gracile nucleus
gr   gracile fasciculus
GrO   granular cell layer of the olfactory bulb
GrI   granular insular cortex
GrC   granular layer of the cochlear nuclei
GrDG   granular layer of the dentate gyrus
GrA   granule cell layer of the accessory olfactory bulb
Gus   gustatory thalamic nucleus
Gus   gustatory thalamic nucleus

H

hbc   habenular commissure
HiI   hilus of the dentate gyrus
hf   hippocampal fissure
12N   hypoglossal nucleus

I

IG   indusium griseum
IO   inferior olive
IOBe   inferior olive, beta subnucleus
IOK   inferior olive, cap of Kooy of the medial nucleus
IOD   inferior olive, dorsal nucleus
IODMC inferior olive, dorsomedial cell column
IODM   inferior olive, dorsomedial cell group
IOM   inferior olive, medial nucleus
IOPr   inferior olive, principal nucleus
IOA   inferior olive, subnucleus A of medial nucleus
IOB   inferior olive, subnucleus B of medial nucleus
IOC   inferior olive, subnucleus C of medial nucleus
IOVL   inferior olive, ventrolateral protrusion
Inf   infracerebellar nucleus
IL   infralimbic cortex
IRe   infratrochlear recess
InfS   infundibular stem
IBl   inner blade of the dentate gyrus
IAD   interanterodorsal thalamic nucleus
IAM   interanteromedial thalamic nucleus
IM   intercalated amygdaloid nucleus, main part
I   intercalated nuclei of the amygdala
In   intercalated nucleus of the medulla
InCo   intercollicular nucleus
icf   interncular fissure
IGL   intergeniculate leaf
ias   intermediate acoustic stria
IntG   intermediate geniculate nucleus
         intermediate interstitial nucleus of the medial longitudinal fasciculus
II
IPit   intermediate lobe of pituitary
ILL   intermediate nucleus of the lateral lemniscus
IMD   interomediodorsal thalamic nucleus
imvc   intermediodorsoventral thalamic commissure
IMn   intermedius nucleus of the medulla
ic   internal capsule
iml   internal medullary lamina
IPl   internal plexiform layer of the olfactory bulb
13   interoculomotor nucleus
IPF   interpeduncular fossa
IP   interpeduncular nucleus
IPA   interpeduncular nucleus, apical subnucleus
IPC   interpeduncular nucleus, caudal subnucleus
IPDM   interpeduncular nucleus, dorsomedial subnucleus
IPI   interpeduncular nucleus, intermediate subnucleus
IPL   interpeduncular nucleus, lateral subnucleus
IPR   interpeduncular nucleus, rostral subnucleus
IPRL   interpeduncular nucleus, rostrolateral subnucleus
IntDM interposed cerebellar nucleus, dorsomedial crest
IntP   interposed cerebellar nucleus, posterior part
IntPPC interposed cerebellar nucleus, posterior parvicellular part
InC   interstitial nucleus of Cajal
InCG   interstitial nucleus of Cajal, greater part
IPACC interstitial nucleus of the posterior limb of the anterior commissure, central part
IPACL interstitial nucleus of the posterior limb of the anterior commissure, lateral part
IPACM interstitial nucleus of the posterior limb of the anterior commissure, medial part
I8   interstitial nucleus of the vestibulocochlear nerve
IVF   interventricular foramen
IMA   intramedullary thalamic area
ICjM   islands of Calleja, major island

J

jx   juxtarestiform body
Jx5   juxtatrigeminal area

K

KF   Kölliker–Fuse nucleus

L

lacunosum moleculare layer of the hippocampus   LMol
lambdoid septal zone   Ld
lamina dissecans of the entorhinal cortex   Dsc
lateral (dentate) cerebellar nucleus   Lat
lateral accumbens shell   LAcbSh
lateral amygdaloid nucleus   La
lateral amygdaloid nucleus, dorsolateral part   LaDL
lateral amygdaloid nucleus, ventrolateral part   LaVL
lateral amygdaloid nucleus, ventromedial part   LaVM
lateral cerebellar nucleus, parvicellular part   LatPC
lateral entorhinal cortex   LEnt
lateral globus pallidus   LGP
lateral habenular nucleus, lateral part   LHbL
lateral habenular nucleus, medial part   LHbM
lateral hypothalamic area   LH
lateral lemniscus   ll
lateral mammillary nucleus   LM
lateral olfactory tract   lo
lateral orbital cortex   LO
lateral parabrachial nucleus, central part   LPBC
lateral parabrachial nucleus, crescent part   LPBCr
lateral parabrachial nucleus, dorsal part   LPBD
lateral parabrachial nucleus, ventral part   LPBV
lateral paragigantocellular nucleus   LPGi
lateral parietal association cortex   LPtA
lateral posterior thalamic nucleus   LP
lateral posterior thalamic nucleus, laterocaudal part   LPLC
lateral posterior thalamic nucleus, laterorostral part   LPLR
lateral posterior thalamic nucleus, mediorostral part   LPMR
lateral preoptic area   LPO
lateral recess of the 4th ventricle   LR4V
lateral reticular nucleus   LRt
lateral reticular nucleus, parvicellular part   LRtPC
lateral reticular nucleus, subtrigeminal part   LRtS5
lateral septal nucleus   LS
lateral septal nucleus, dorsal part   LSD
lateral septal nucleus, intermediate part   LSI
lateral septal nucleus, ventral part   LSV
lateral stripe of the striatum   LSS
lateral superior olive   LSO
lateral tegmental tract   ltg
lateral terminal nucleus of the accessory optic tract   LT
lateral ventricle   LV
lateral vestibular nucleus   LVe
lateral vestibulospinal tract   lvs
lateroanterior hypothalamic nucleus   LA
laterodorsal tegmental nucleus   LDTg
laterodorsal tegmental nucleus, ventral part   LDTgV
laterodorsal thalamic nucleus   LD
laterodorsal thalamic nucleus, dorsomedial part   LDDM
laterodorsal thalamic nucleus, ventrolateral part   LDVL
lateroventral periolivary nucleus   LVPO
linear nucleus of the medulla   Li
locus coeruleus   LC
longitudinal association bundle   lab

M

magnocellular nucleus of the lateral hypothalamus   MCLH
magnocellular nucleus of the posterior commissure   MCPC
magnocellular preoptic nucleus   MCPO
mammillary peduncle   mp
mammillary recess of the 3rd ventricle   MRe
mammillotegmental tract   mtg
mammillothalamic tract   mt
marginal zone of the medial geniculate   MZMG
medial (fastigial) cerebellar nucleus   Med
medial accessory oculomotor nucleus   MA3
medial amygdaloid nucleus   Me
medial amygdaloid nucleus, anterior dorsal   MeAD
medial amygdaloid nucleus, anterior part   MeA
medial amygdaloid nucleus, anteroventral part   MeAV
medial amygdaloid nucleus, posterodorsal part   MePD
medial cerebellar nucleus, dorsolateral protuberance   MedDL
medial corticohypothalamic tract   mch
medial eminence, external layer   MEE
medial eminence, internal layer   MEI
medial entorhinal cortex   MEnt
medial entorhinal cortex, ventral part   MEntV
medial forebrain bundle   mfb
medial geniculate nucleus   MG
medial geniculate nucleus, dorsal part   MGD
medial geniculate nucleus, medial part   MGM
medial globus pallidus (entopeduncular nucleus)   MGP
medial habenular nucleus   MHb
medial lemniscus   ml
medial longitudinal fasciculus   mlf
medial mammillary nucleus, lateral part   ML
medial orbital cortex   MO
medial parabrachial nucleus   MPB
medial parabrachial nucleus external part   MPBE
medial parietal association cortex   MPtA
medial pontine nucleus   MPn
medial preoptic area   MPA
medial preoptic nucleus   MPO
medial preoptic nucleus, central part   MPOC
medial preoptic nucleus, lateral part   MPOL
medial pretectal nucleus   MPT
medial rostroventrolateral medulla   MRVL
medial septal nucleus   MS
medial superior olive   MSO
medial terminal nucleus of the accessory optic tract   MT
medial tuberal nucleus   MTu
medial vestibular nucleus   MVe
medial vestibular nucleus, magnocellular part   MVeMC
medial vestibular nucleus, parvicellular part   MVePC
median accessory nucleus of the medulla   MnA
median raphe nucleus   MnR
mediodorsal thalamic nucleus, central part   MDC
mediodorsal thalamic nucleus, lateral part   MDL
mediodorsal thalamic nucleus, medial part   MDM
mediodorsal thalamic nucleus, paralaminar part   MDPL
medullary reticular nucleus, dorsal part   MdD
medullary reticular nucleus, ventral part   MdV
mesencephalic trigeminal nucleus   Me5
mesencephalic trigeminal tract   me5
microcellular tegmental nucleus   MiTg
middle cerebellar peduncle   mcp
middle cerebral artery   mcer
minimus nucleus   Min
mitral cell layer of the accessory olfactory bulb   MiA
mitral cell layer of the olfactory bulb   Mi
molecular layer of the dentate gyrus   Mol
motor root of the trigeminal nerve   m5
motor trigeminal nucleus   Mo5
motor trigeminal nucleus, dorsolateral part   Mo5DL
motor trigeminal nucleus, ventromedial part   Mo5VM

N

nigrostriatal bundle   ns
noradrenaline/adrenaline cells   A1/C1
nucleus O   O
nucleus of Darkschewitsch   Dk
nucleus of origin of efferents of the vestibular nerve   EVe
nucleus of the brachium of the inferior colliculus   BIC
nucleus of the commissural stria terminalis   CST
nucleus of the dorsal hippocampal commissure   DHC
nucleus of the fields of Forel   F
nucleus of the horizontal limb of the diagonal band   HDB
nucleus of the optic tract   OT
nucleus of the posterior commissure   PCom
nucleus of the solitary tract   Sol
nucleus of the solitary tract, central part   SolCe
nucleus of the solitary tract, commissural part   SolC
nucleus of the solitary tract, dorsomedial part   SolDM
nucleus of the solitary tract, intermediate part   SolIM
nucleus of the solitary tract, interstitial part   SolI
nucleus of the solitary tract, medial part   SolM
nucleus of the solitary tract, rostrolateral part 1–10   SolRL
nucleus of the solitary tract, ventrolateral part   SolVL
nucleus of the solotary tract, rostral central part   RCe
nucleus of the stria medullaris   SM
nucleus of the trapezoid body   Tz
nucleus of the vertical limb of the diagonal band   VDB
nucleus X   X
nucleus Y   Y
nucleus Z   Z

O

obex   Obex
olfactory nerve layer   ON
olfactory tubercle   Tu
olfactory tubercle densocellular layer   TuDC
olfactory tubercle plexiform layer   TuPl
olfactory tubercle polymorph layer   TuPo
olfactory ventricle (olfactory part of lateral ventricle)   OV
olivary pretectal nucleus   OPT
olivocerebellar tract   oc
olivocochlear bundle   ocb
optic chiasm   ox
optic nerve   2n
optic nerve layer of the superior colliculus   Op
oriens layer of the hippocampus   Or
outer blade of the dentate gyrus   OBl
oval paracentral thalamic nucleus   OPC

P

parabigeminal nucleus   PBG
parabrachial nucleus   PB
parabrachial nucleus, waist part   PBW
parabrachial pigmented nucleus   PBP
paracentral thalamic nucleus   PC
paracochlear glial substance   PCGS
paracollicular tegmentum   PCTg
parafascicular thalamic nucleus   PF
paraflocular sulcus   pfs
paraflocculus   PFl
paralemniscal nucleus   PL
paramedian lobule   PM
paramedian raphe nucleus   PMnR
paramedian reticular nucleus   PMn
paramedian sulcus   pms
paranigral nucleus   PN
pararubral nucleus   PaR
parasolitary nucleus   PSol
parastrial nucleus   PS
parasubiculum   PaS
parasubthalamic nucleus   PSTh
paratrigeminal nucleus   Pa5
paraventricular hypothalamic nucleus, anterior
      magnocellular part   PaAM
paraventricular hypothalamic nucleus, anterior
      parvicellular part   PaAP
paraventricular hypothalamic nucleus, dorsal cap
      PaDC
paraventricular hypothalamic nucleus, lateral
      magnocellular part   PaLM
paraventricular hypothalamic nucleus, medial
      parvicellular part   PaMP
paraventricular hypothalamic nucleus, posterior part
      PaPo
paraventricular hypothalamic nucleus, ventral part
      PaV
paraventricular thalamic nucleus   PV
paraventricular thalamic nucleus, anterior part   PVA
paraventricular thalamic nucleus, posterior part   PVP
parvicellular motor trigeminal nucleus   PC5
parvicellular reticular nucleus   PCRt
pedunculopontine tegmental nucleus   PPTg
periaqueductal gray   PAG
perifacial zone   P7
perifornical nucleus   PeF
peripeduncular nucleus   PP
perirhinal cortex   PRh
peritrigeminal zone   P5

periventricular fiber system   pv
periventricular hypothalamic nucleus   Pe
pineal gland   Pi
pineal recess   PiRe
piriform cortex   Pir
polymorph layer of the dentate gyrus   PoDG
pontine nuclei   Pn
pontine raphe nucleus   PnR
pontine reticular nucleus, caudal part   PnC
pontine reticular nucleus, oral part   PnO
pontine reticular nucleus, ventral part   PnV
posterior commissure   pc
posterior hypothalamic area   PH
posterior intralaminar thalamic nucleus   PIL
posterior limitans thalamic nucleus   PLi
posterior pretectal nucleus   PPT
posterior superior fissure   psf
posterior thalamic nuclear group   Po
posterior thalamic nuclear group, triangular part   PoT
posterodorsal preoptic nucleus   PDP
posterodorsal tegmental nucleus   PDTg
posterolateral cortical amygdaloid nucleus (C2)   PLCo
posterolateral fissure   plf
posteromedial cortical amygdaloid nucleus (C3)
      PMCo
postsubiculum   Post
pre-Bötzinger complex   PrBo
precentral fissure   pcn
precommissural fornix   pcf
preculminate fissure   pcuf
predorsal bundle   pd
prelimbic cortex   PrL
premammillary nucleus, dorsal part   PMD
premammillary nucleus, ventral part   PMV
prepositus nucleus   Pr
prepyramidal fissure   ppf
prerubral field   PR
presubiculum   PrS
primary auditory cortex   Au1
primary fissure   prf
primary motor cortex   M1
primary somatosensory cortex   S1
primary somatosensory cortex, barrel field   S1BF
primary somatosensory cortex, dysgranular region
      S1DZ
primary somatosensory cortex, forelimb region   S1FL
primary somatosensory cortex, hindlimb region   S1HL
primary somatosensory cortex, jaw region   S1J
primary somatosensory cortex, jaw region, oral surface
      S1JO

primary somatosensory cortex, shoulder region   S1Sh
primary somatosensory cortex, upper lip region
      S1ULp
primary visual cortex   V1
primary visual cortex, binocular area   V1B
primary visual cortex, monocular area   V1M
principal mammillary tract   pm
principal sensory trigeminal nucleus   Pr5
principal sensory trigeminal nucleus, dorsomedial part
      Pr5DM
pyramidal decussation   pyx
pyramidal fissure   pf
pyramidal tract   py

R

raphe cap   RC
raphe interpositus nucleus   RIP
raphe magnus nucleus   RMg
raphe obscurus nucleus   ROb
raphe pallidus nucleus   RPa
recess of the inferior colliculus   ReIC
red nucleus   R
red nucleus, lateral horn   RLH
red nucleus, magnocellular part   RMC
reticular thalamic nucleus   Rt
reticulotegmental nucleus of the pons   RtTg
reticulotegmental nucleus of the pons, pericentral part
      RtTgP
retroambiguus nucleus   RAmb
retrochiasmatic area   RCh
retroethmoid nucleus   REth
retrolemniscal nucleus   RL
retroparafascicular nucleus   RPF
retrorubral field   RRF
retrorubral fields/A8 dopamine cells   RRF/A8
retrorubral nucleus   RR
retrosplenial agranular cortex   RSA
retrosplenial granular a cortex   RSGa
retrosplenial granular b cortex   RSGb
retrosplenial granular cortex   RSG
reuniens thalamic nucleus   Re
rhinal fissure   rf
rhinal incisura   ri
rhomboid thalamic nucleus   Rh
root of abducens nerve   6n
root of accessory nerve   11n
root of hypoglossal nerve   12n
rostral interstitial nucleus of medial longitudinal
      fasciculus   RI

rostral linear nucleus of the raphe   RLi
rostral periolivary region   RPO
rostroventrolateral reticular nucleus   RVL
rostrum of the corpus callosum   rcc
rubrospinal tract   rs

S

sagulum nucleus   Sag
scaphoid thalamic nucleus   Sc
secondary auditory cortex, dorsal area   AuD
secondary auditory cortex, ventral area   AuV
secondary motor cortex   M2
secondary somatosensory cortex   S2
secondary visual cortex, lateral area   V2L
secondary visual cortex, mediolateral area   V2ML
semilunar nucleus   SL
sensory root of the trigeminal nerve   s5
septofimbrial nucleus   SFi
septohippocampal nucleus   SHi
simple lobule   Sim
simple lobule A   SimA
solitary tract   sol
sphenoid nucleus   Sph
spinal trigeminal nucleus, caudal part   Sp5C
spinal trigeminal nucleus, oral part   Sp5O
spinal trigeminal nucleus, oral part, ventrolateral
      division   Sp5OVL
spinal trigeminal tract   sp5
spinal vestibular nucleus   SpVe
spinocerebellar tract   sc
splenium of the corpus callosum   scc
stratum lucidum, hippocampus   SLu
stria terminalis   st
striohypothalamic nucleus   StHy
subcoeruleus nucleus, alpha part   SubCA
subcoeruleus nucleus, dorsal part   SubCD
subcoeruleus nucleus, ventral part   SubCV
subcommissural nucleus   SCom
subcommissural organ   SCO
subfornical organ   SFO
subgeniculate nucleus   SubG
subiculum   S
subincertal nucleus   SubI
sublenticular extended amygdala   SLEA
sublenticular extended amygdala, central part
      SLEAC
sublenticular extended amygdala, medial part
      SLEAM
submammillothalamic nucleus   SMT

submedius thalamic nucleus   Sub
submedius thalamic nucleus, dorsal part   SubD
submedius thalamic nucleus, ventral part   SubV
subparafascicular thalamic nucleus   SPF
subparafascicular thalamic nucleus, parvicellular part   SPFPC
subparaventricular zone of the hypothalamus   SPa
subpeduncular tegmental nucleus   SPTg
substantia innominata   SI
substantia innominata, basal part   SIB
substantia innominata, dorsal part   SID
substantia innominata, ventral part   SIV
substantia nigra   SN
substantia nigra, compact part   SNC
substantia nigra, compact part, dorsal tier   SNCD
substantia nigra, compacta part, ventral tier   SNCV
substantia nigra, lateral part   SNL
substantia nigra, medial part   SNM
substantia nigra, reticular part   SNR
substantia nigra, reticular part, dorsomedial tier   SNRDM
substantia nigra, reticular part, ventrolateral tier   SNRVL
subthalamic nucleus   STh
sulcus limitans   slim
superficial glial zone of the cochlear nuclei   SGl
superficial gray layer of the superior colliculus   SuG
superior cerebellar peduncle (brachium conjunctivum)   scp
superior cerebellar peduncle, descending limb   scpd
superior paraolivary nucleus   SPO
superior salivatory nucleus   SuS
superior thalamic radiation   str
superior vestibular nucleus   SuVe
suprachiasmatic nucleus   SCh
suprachiasmatic nucleus, dorsomedial part   SChDM
suprachiasmatic nucleus, ventrolateral part   SChVL
suprageniculate thalamic nucleus   SG
supragenual nucleus   SGe
supramammillary decussation   sumx
supramammillary nucleus   SuM
supramammillary nucleus, medial part   SuMM
supraoculomotor cap   Su3C
supraoculomotor periaqueductal gray   Su3
supraoptic decussation   sox
supraoptic nucleus   SO
supratrigeminal nucleus   Su5

T

tectospinal tract   ts
terete hypothalamic nucleus   Te
trigeminal ganglion   5Gn
trigeminal ganglion   5Gn
trochlear decussation   4x
trochlear nucleus   4N

U

uncinate fasciculus   unc
uvula, cerebellum   Uvu
uvular fissure   uf

V

vagus nerve or its root   10n
ventral anterior thalamic nucleus   VA
ventral cochlear nucleus   VC
ventral cochlear nucleus, anterior part   VCA
ventral cochlear nucleus, posterior part   VCP
ventral hippocampal commissure   vhc
ventral lateral geniculate nucleus   VLG
ventral lateral geniculate nucleus, magnocellular part   VLGMC
ventral lateral geniculate nucleus, parvicellular part   VLGPC
ventral nucleus of the lateral lemniscus   VLL
ventral orbital cortex   VO
ventral pallidum   VP
ventral peduncular pontine nucleus   VPPn
ventral posterolateral thalamic nucleus   VPL
ventral posteromedial thalamic nucleus   VPM
ventral reuniens thalamic nucleus   VRe
ventral spinocerebellar tract   vsc
ventral tegmental area   VTA
ventral tegmental decussation   vtgx
ventral tenia tecta   VTT
ventral tuberomammillary nucleus   VTM
ventrolateral hypothalamic nucleus   VLH
ventrolateral preoptic nucleus   VLPO
ventrolateral thalamic nucleus   VL
ventromedial hypothalamic nucleus   VMH
ventromedial hypothalamic nucleus, anterior part   VMHA
ventromedial hypothalamic nucleus, central part   VMHC
ventromedial hypothalamic nucleus, dorsomedial part   VMHDM
ventromedial hypothalamic nucleus, ventrolateral part   VMHVL
ventromedial preoptic nucleus   VMPO
ventromedial thalamic nucleus   VM
vertebral artery   vert
vestibular root of the vestibulocochlear nerve   8vn
vestibulocerebellar nucleus   VeCb
vestibulocochlear ganglion   8Gn
vestibulocochlear nerve   8n
vestibulomesencephalic tract   veme
visual tegmental relay zone   VTRZ
vomeronasal nerve   vn
vomeronasal nerve layer   VN

X

xiphoid thalamic nucleus   Xi

Z

zona incerta, dorsal part   ZID
zona incerta, ventral part   ZIV
zona limitans   ZL
zonal layer of the superior colliculus   Gl

Index of Abbreviations

The abbreviations are listed in alphabetical order. Each abbreviation is followed by the structure name and the number of the figures on which the abbreviation appears.

1   layer 1   10–18, 35–38, 70–74, 102
1Cb   1st Cerebellar lobule   79–80, 101
2   layer 2   10–18, 27, 35–38, 70–74
2Cb   2nd Cerebellar lobule   72–80, 101–111
2n   optic nerve   30
3   layer 3   10–14, 35–38, 70–74
3Cb   3rd Cerebellar lobule   74–81, 101–117, 119
3N   oculomotor nucleus   62–66, 101–103
3n   oculomotor nerve or its root   58, 60–61
3PC   oculomotor nucleus, parvicellular part   62–66, 102–103
3V   3rd ventricle   27–52, 101–107
4&5Cb   4&5th Cerebellar lobules   75–87, 101–112, 119, 121
4N   trochlear nucleus   68, 102–103
4n   trochlear nerve or its root   70–75, 108–110, 112
4V   4th ventricle   74–91, 101, 103–108
6   abducens nucleus   86
6Cb   6th Cerebellar lobule   85, 87–94, 101–114
6N   abducens nucleus   77–79, 103–105
7Cb   7th Cerebellar lobule   91–98, 101–113
7DI   facial nucleus, dorsal intermediate subnucleus   81
7DL   facial nucleus, dorsolateral subnucleus   81
7DM   facial nucleus, dorsomedial subnucleus   81
7L   facial nucleus, lateral subnucleus   81
7N   facial nucleus   78–85, 107–115
7n   facial nerve or its root   74–78, 106–115
7VI   facial nucleus, ventral intermediate subnucleus   81
7VM   facial nucleus, ventromedial subnucleus   81
8Cb   8th Cerebellar lobule   89–102, 104–114
8n   vestibulocochlear nerve   74–82, 115–121
8vn   vestibular root of the vestibulocochlear nerve   72–73, 78–80
9Cb   9th Cerebellar lobule   87–110
10   dorsal motor nucleus of vagus   87
10Cb   10th Cerebellar lobule   81–86, 88–89, 101–113

10N   dorsal motor nucleus of vagus   84–99, 101–104
11N   accessory nerve nucleus   100
11n   root of accessory nerve   92
12   hypoglossal nucleus   91
12N   hypoglossal nucleus   89–90, 92–99, 101–103
12n   root of hypoglossal nerve   91, 96–98

A

A1   A1 noradrenaline cells   93–96, 100
A1/C1   noradrenaline/adrenaline cells   97
A11   A11dopamine cells   47
A12   A12 dopamine cells   47–48
A13   A13 dopamine cells   40–43, 106–108
A14   A14 dopamine cells   32–34
A2   A2 noradrenaline cells   100
A5   A5 noradrenaline cells   75–81, 112, 114
A7   A7 noradrenaline cells   72, 115
A8   A8 dopamine cells   63–65
AAD   anterior amygdaloid area, dorsal part   32–38, 115–119
AAV   anterior amygdaloid area, ventral part   32–36, 114–119
AC   anterior commissural nucleus   33–35, 102
ac   anterior commissure   101–109
aca   anterior commissure, anterior part   14–32, 108–113
Acb   accumbens nucleus   14, 25–26
AcbC   accumbens nucleus, core   15–24, 106–113
AcbSh   accumbens nucleus, shell   15–24, 103–112
aci   anterior commissure, intrabulbar part   1–13, 107–109, 111
ACo   anterior cortical amygdaloid nucleus   33–45, 119–124
acp   anterior commissure, posterior   29–35, 110–123
Acs   accessory nucleus of the ventral horn   37
Acs5   accessory trigeminal nucleus   73–75, 109
Acs6/7   accessory abducens/facial nucleus   77
Acs7   accessory facial nucleus   78–82
AD   anterodorsal thalamic nucleus   34–40, 106–109
ADP   anterodorsal preoptic nucleus   28–34, 102–103
AHA   anterior hypothalamic area, anterior part   34–36, 103–106
AHC   anterior hypothalamic area, central part   37–38, 102–106
AHi   amygdalohippocampal area   119–124
AHiAL   amygdalohippocampal area, anterolateral part   47–52
AHiPM   amygdalohippocampal area, posteromedial part   51–62, 114–118

AHP   anterior hypothalamic area, posterior part   39–42, 103–106
AI   agranular insular cortex   0, 10–14
AID   agranular insular cortex, dorsal part   0, 15–27, 114–132
AIP   agranular insular cortex, posterior part   0, 28–40, 129–132
AIV   agranular insular cortex, ventral part   0, 15–27, 114–132
AL   nucleus of the ansa lenticularis   40
al   ansa lenticularis   41
alv   alveus of the hippocampus   40–67, 105–123, 125–126, 128–130, 132
AM   anteromedial thalamic nucleus   35–41, 104–108
Amb   ambiguus nucleus   86–97, 111–112
AMV   anteromedial thalamic nucleus, ventral part   36–38, 105–106
Ang   angular thalamic nucleus   43–44, 109–111
Ant   anterior lobe cerebellum   71–75
AOB   accessory olfactory bulb   4–6, 106–107, 109, 111–112
AOD   anterior olfactory nucleus, dorsal part   4–10, 108–111
AOE   anterior olfactory nucleus, external part   3–6, 106–113
AOL   anterior olfactory nucleus, lateral part   3–10, 111–113
AOM   anterior olfactory nucleus, medial part   4–14, 104–107
AOP   anterior olfactory nucleus, posterior part   11–14, 105–111
aot   accessory olfactory tract   42
AOV   anterior olfactory nucleus, ventral part   4–10, 105–111
AP   area postrema   91–96
APir   amygdalopiriform transition area   51–62, 125–132
APit   anterior lobe of pituitary   107
apmf   ansoparamedian fissure   85–95, 114–129
APT   anterior pretectal nucleus   56–59, 107–108, 112–115
APTD   anterior pretectal nucleus, dorsal part   49–55, 109–111
APTV   anterior pretectal nucleus, ventral part   53–55, 110–111
Aq   aqueduct (Sylvius)   55–73, 101–102
Arc   arcuate hypothalamic nucleus   41–44, 102–103
ArcD   arcuate hypothalamic nucleus, dorsal part   45–48

ArcL   arcuate hypothalamic nucleus, lateral part   45–48
ArcLP   arcuate hypothalamic nucleus, lateroposterior part   49–52
ArcMP   arcuate hypothalamic nucleus, medial posterior part   49–54, 101
asc7   ascending fibers of the facial nerve   105–106
AStr   amygdalostriatal transition area   39–49, 124–130
ATg   anterior tegmental nucleus   65–66, 68–69, 102–103
Au1   primary auditory cortex   0, 49–61, 131–132
AuD   secondary auditory cortex, dorsal area   0, 48–61, 129–132
AuV   secondary auditory cortex, ventral area   0, 45–59, 131–132
AV   anteroventral thalamic nucleus   34–35, 39–41
AVDM   anteroventral thalamic nucleus, dorsomedial part   36–38, 107–110
AVPe   anteroventral periventricular nucleus   26–31, 102
AVVL   anteroventral thalamic nucleus, ventrolateral part   36–38, 108–112

B

B   basal nucleus (Meynert)   34–40, 42, 113–121
B4   B4 serotonin cells   88–89
B9   B9 serotonin cells   64–66, 68
BAC   bed nucleus of the anterior commissure   31–32, 105
BAOT   bed nucleus of the accessory olfactory tract   38–39, 116–117
Bar   Barrington's nucleus   75–78, 107
bas   basilar artery   61
BIC   nucleus of the brachium of the inferior colliculus   62–69, 115–118
bic   brachium of the inferior colliculus   59–69, 118–119
BL   basolateral amygdaloid nucleus   121
BLA   basolateral amygdaloid nucleus, anterior part   36–48, 124–128
BLP   basolateral amygdaloid nucleus, posterior part   47–57, 122–130
BLV   basolateral amygdaloid nucleus, ventral part   40–49, 127–130
BMA   basomedial amygdaloid nucleus, anterior part   36–45, 119–126

BMP   basomedial amygdaloid nucleus, posterior part   43–53, 120–128
Bo   Botzinger complex   86
bp   brachium pontis (stem of middle cerebellar peduncle)   62–65
bsc   brachium of the superior colliculus   52–66, 105–121
BST   bed nucleus of the stria terminalis   26–27
BSTIA   bed nucleus of the stria terminalis, intraamygdaloid division   42–49, 121
BSTLD   bed nucleus of the stria terminalis, lateral division, dorsal part   28–31, 107–110
BSTLI   bed nucleus of the stria terminalis, lateral division, intermediate part   31–32, 108
BSTLJ   bed nucleus of the stria terminalis, lateral division, juxtacapsular part   29–31, 110
BSTLP   bed nucleus of the stria terminalis, lateral division, posterior part   29–33, 107–110
BSTLV   bed nucleus of the stria terminalis, lateral division, ventral part   27–31, 107–109
BSTMA   bed nucleus of the stria terminalis, medial division, anterior part   26–31, 105–106
BSTMP   bed nucleus of the stria terminalis, medial division, posterior part   105
BSTMPI   bed nucleus of the stria terminalis, medial division, posterointermediate part   32–34, 106–108
BSTMPL   bed nucleus of the stria terminalis, medial division, posterolateral part   33–36, 106–110
BSTMPM   bed nucleus of the stria terminalis, medial division, posteromedial part   32–34, 104–108
BSTMV   bed nucleus of the stria terminalis, medial division, ventral part   28–32, 107
BSTS   bed nucleus of stria terminalis, supracapsular part   31, 33–39

C

C1   C1 adrenaline cells   84–89, 98, 107–115
C2   C2 adrenaline cells   88–89
C3   C3 adrenaline cells   86–89
CA1   field CA1 of hippocampus   41–63, 101–117, 119, 121–132
CA2   field CA2 of hippocampus   41–56, 104–132
CA3   field CA3 of hippocampus   39–61, 105–106, 108–132
CB   cell bridges of the ventral striatum   23–24
Cb   cerebellum   86–87
cbc   cerebellar commissure   84–86
CC   central canal   93–100
cc   corpus callosum   25–51, 101–129
CeC   central amygdaloid nucleus, capsular part   38–46, 120–125
CeCv   central cervical nucleus   97–100
CeL   central amygdaloid nucleus, lateral division   41–48, 122–124
CeM   central amygdaloid nucleus, medial division   36–38, 41, 118–122
CeMAD   central amygdaloid nucleus, medial division, anterodorsal part   39–40
CeMAV   central amygdaloid nucleus, medial division, anteroventral part   39–40
CeMPV   central amygdaloid nucleus, medial posteroventral part   42–44
cg   cingulum   18–63, 107–110
Cg/RS   cingulate/retrosplenial cortex   0, 34–38, 101–105
Cg1   cingulate cortex, area 1   0, 11–33, 101–106
Cg2   cingulate cortex, area 2   0, 19–33, 101–105
CGA   central gray, alpha part   75–77, 101–102
CGB   central gray, beta part   75
CGPn   central gray of the pons   76–79, 104–106
chp   choroid plexus   102, 105–106, 118, 120, 122–123, 125
CI   caudal interstitial nucleus of the medial longitudinal fasciculus   87–90
CIC   central nucleus of the inferior colliculus   72–75, 106–114
cic   commissure of the inferior colliculus   71–72, 101–105
Cir   circular nucleus   37–39
CL   centrolateral thalamic nucleus   40–48, 107–109
Cl   claustrum   14–40, 106–131
CLi   caudal linear nucleus of the raphe   64–66, 68, 101–103
CM   central medial thalamic nucleus   35–48, 101–105
CnF   cuneiform nucleus   68–74, 109–114
Cop   copula of the pyramis   87–98, 115–123
cp   cerebral peduncle, basal part   46–65, 108–120
CPO   caudal periolivary nucleus   78
CPu   caudate putamen (striatum)   15–50, 106–131
Crus1   crus 1 of the ansiform lobule   76–91, 116–131
Crus2   crus 2 of the ansiform lobule   80–95, 115–132
csc   commissure of the superior colliculus   56–60, 101–102
cst   commissural stria terminalis   73
Cu   cuneate nucleus   91–100, 104–111
cu   cuneate fasciculus   96–100, 104–108
CVL   caudoventrolateral reticular nucleus   110–113
CxA   cortex-amygdala transition zone   32–42, 120–126

D

D3V   dorsal 3rd ventricle   31–54, 101–102, 104–105
das   dorsal acoustic stria   84
DC   dorsal cochlear nucleus   78–85, 114–121
DCIC   dorsal cortex of the inferior colliculus   72–76, 101–109
DEn   dorsal endopiriform nucleus   10–54, 110–132
df   dorsal fornix   33–50, 101–102, 104
DG   dentate gyrus   39–64, 102–103, 113, 118–119, 121
dhc   dorsal hippocampal commissure   39–71, 101–102
DI   dysgranular insular cortex   15–40, 120–132
Dk   nucleus of Darkschewitsch   54–61, 102–104
DLG   dorsal lateral geniculate nucleus   45–55, 118–122
DLL   dorsal nucleus of the lateral lemniscus   69–72, 116–118
DLO   dorsolateral orbital cortex   0, 6–9
dlo   dorsal lateral olfactory tract   3–5, 108, 110, 112–113
DLPAG   dorsolateral periaqueductal gray   59–72, 104–107
DM   dorsomedial hypothalamic nucleus   42–45, 48–49, 102–106
DMC   dorsomedial hypothalamic nucleus, compact part   46–48
DMD   dorsomedial hypothalamic nucleus, dorsal part   46–47
DMPAG   dorsomedial periaqueductal gray   59–74, 101–103
DMPn   dorsomedial pontine nucleus   66, 68
DMSp5   dorsomedial spinal trigeminal nucleus   82–91, 113–115
DMTg   dorsomedial tegmental area   72–76, 103–107
DMV   dorsomedial hypothalamic nucleus, ventral part   46–47
DP   dorsal peduncular cortex   0, 13–23, 102–106
DpG   deep gray layer of the superior colliculus   56–71, 103–113
DPGi   dorsal paragigantocellular nucleus   81–89, 101–104
DpMe   deep mesencephalic nucleus   55–67, 105, 107–115
DPO   dorsal periolivary region   73–77, 107–113
DpWh   deep white layer of the superior colliculus   56–71, 103–111
DR   dorsal raphe nucleus   64–66, 102
DRC   dorsal raphe nucleus, caudal part   72–74
DRD   dorsal raphe nucleus, dorsal part   68–71, 101
DRI   dorsal raphe nucleus, interfascicular part   68–75, 101
DRV   dorsal raphe nucleus, ventral part   68–71, 101, 103–105
DRVL   dorsal raphe nucleus, ventrolateral part   68–71, 103
Dsc   lamina dissecans of the entorhinal cortex   69, 126–132
dsc   dorsal spinocerebellar tract   97, 99–100
DT   dorsal terminal nucleus of the accessory optic tract   58–61, 116
dtg   dorsal tegmental bundle   102
DTgC   dorsal tegmental nucleus, central part   74–77, 101–103
DTgP   dorsal tegmental nucleus, pericentral part   73–77, 101–103
dtgx   dorsal tegmental decussation   63, 101
DTM   dorsal tuberomammillary nucleus   49–52, 102
DTr   dorsal transition zone   103
DTT   dorsal tenia tecta   9–20, 101–107

E

E   ependyma and subependymal layer   108
E/OV   ependymal and subendymal layer/olfactory ventricle   1–11, 106, 108
ec   external capsule   18–67, 127–128, 130
ECIC   external cortex of the inferior colliculus   65–66, 68–75, 106–117
Ect   ectorhinal cortex   0, 42–74, 129–132
ECu   external cuneate nucleus   87–95, 112
EF   epifascicular nucleus   88
EMi   epimicrocellular nucleus   66, 68–70
eml   external medullary lamina   45–50, 114
Ent   entorhinal cortex   128
EPl   external plexiform layer of the olfactory bulb   1–9, 101–115
EPlA   external plexiform layer of the accessory olfactory bulb   3, 5, 108
ERS   epirubrospinal nucleus   114
Eth   ethmoid thalamic nucleus   52–54, 110–113
EVe   nucleus of origin of efferents of the vestibular nerve   78–79
EW   Edinger-Westphal nucleus   54–64, 101
exc   extreme capsule   29–32

F

F   nucleus of the fields of Forel   51–53, 108–110

f   fornix   30–56, 101–108

FC   fasciola cinereum   45–54

FF   fields of Forel   50, 107

fi   fimbria of the hippocampus   34–52, 106–120, 122–126

Fl   flocculus   73–82, 122–127

fmi   forceps minor of the corpus callosum   14–20, 109–124

fmj   forceps major of the corpus callosum   53–56, 108–124

fr   fasciculus retroflexus   42–57, 103–105

FrA   frontal association cortex   0, 4–9, 104–116

Fu   bed nucleus of stria terminalis, fusiform part   29–30, 107–109

FVe   F cell group of the vestibular complex   87–89

G

g7   genu of the facial nerve   79–80, 104–105

gcc   genu of the corpus callosum   21–24, 101–106

Ge5   gelatinous layer of the caudal spinal trigeminal nucleus   111, 113, 115

Gem   gemini hypothalamic nucleus   49–51, 107

GI   granular insular cortex   18–40, 121–132

Gi   gigantocellular reticular nucleus   78–91, 101–108

GiA   gigantocellular reticular nucleus, alpha part   76–85, 101–106

GiV   gigantocellular reticular nucleus, ventral part   86–89, 101–106

Gl   glomerular layer of the olfactory bulb   1–8, 101–116

GlA   glomerular layer of the accessory olfactory bulb   3, 5, 108, 110

Gr   gracile nucleus   95–103

gr   gracile fasciculus   97, 101–102

GrA   granule cell layer of the accessory olfactory bulb   2–3, 108–110

GrC   granular layer of the cochlear nuclei   78–83

GrDG   granular layer of the dentate gyrus   39–64, 104, 106–113, 115–124

GrO   granular cell layer of the olfactory bulb   1–9, 102–113

Gus   gustatory thalamic nucleus   47–50, 102–110

H

hbc   habenular commissure   46–50

HDB   nucleus of the horizontal limb of the diagonal band   24–35, 104–111

I

hf   hippocampal fissure   42–63, 103–110, 112–128

I   intercalated nuclei of the amygdala   28–38, 40, 42–47, 120–124

I3   interoculomotor nucleus   64

I5   intertrigeminal nucleus   73–75, 115

IAD   interanterodorsal thalamic nucleus   36–39, 104–106

IAM   interanteromedial thalamic nucleus   37–41, 101–103

ias   intermediate acoustic stria   83

ic   internal capsule   29–49, 110–127

icf   intercrural fissure   80–91, 114–131

ICj   islands of Calleja   15–21, 23, 25, 27, 101–102, 108, 110, 120

ICjM   islands of Calleja, major island   19–24, 104–106

icp   inferior cerebellar peduncle (restiform body)   78–94, 115–116, 118–119

IF   interfascicular nucleus   56–62, 101–102

IG   indusium griseum   21–51, 101–103

IGL   intergeniculate leaf   49–55, 120–121

IL   infralimbic cortex   14–20, 103–105

ILL   intermediate nucleus of the lateral lemniscus   68–71, 116–118

IM   intercalated amygdaloid nucleus, main part   38–39

IMA   intramedullary thalamic area   48–54

IMD   intermediodorsal thalamic nucleus   41–48, 101

In   intercalated nucleus of the medulla   89–93, 101–102

InC   interstitial nucleus of Cajal   55–62, 103–106

InCG   interstitial nucleus of Cajal, greater part   55–59

InCo   intercollicular nucleus   65–70, 109–114

Inf   infracerebellar nucleus   81

InG   intermediate gray layer of the superior colliculus   56–71, 103–115

InM   intermedius nucleus of the medulla   94–95

IntA   interposed cerebellar nucleus, anterior part   79–84, 109–116

IntDL   interposed cerebellar nucleus, dorsolateral hump   81–85

IntG   intermediate geniculate nucleus   54

IntP   interposed cerebellar nucleus, posterior part   83–87, 108–116

InWh   intermediate white layer of the superior colliculus   56–71, 103–116

IO   inferior olive   101–102, 105–108

IOA   inferior olive, subnucleus A of medial nucleus   94–97

IOB   inferior olive, subnucleus B of medial nucleus   92–97

IOBe   inferior olive, beta subnucleus   91–96

IOC   inferior olive, subnucleus C of medial nucleus   92–97

IOD   inferior olive, dorsal nucleus   87–95, 103–104

IODM   inferior olive, dorsomedial cell group   88–90

IODMC   inferior olive, dorsomedial cell column   89

IOK   inferior olive, cap of Kooy of the medial nucleus   91–96

IOM   inferior olive, medial nucleus   87–91, 98

IOPr   inferior olive, principal nucleus   87–93, 103–104

IOVL   inferior olive, ventrolateral protrusion   91–92

IP   interpeduncular nucleus   65–66

IPA   interpeduncular nucleus, apical subnucleus   64, 101–102

IPAC   interstitial nucleus of the posterior limb of the anterior commissure   37–43, 110–121

IPACL   interstitial nucleus of the posterior limb of the anterior commissure, lateral part   27–36

IPACM   interstitial nucleus of the posterior limb of the anterior commissure, medial part   27–36

IPC   interpeduncular nucleus, caudal subnucleus   59–64, 101–102

IPDL   interpeduncular nucleus, dorsolateral subnucleus   59–64

IPDM   interpeduncular nucleus, dorsomedial subnucleus   59–64

IPF   interpeduncular fossa   56–59, 101–106

IPI   interpeduncular nucleus, intermediate subnucleus   59–64

IPL   interpeduncular nucleus, lateral subnucleus   59–64, 103–104

IPl   internal plexiform layer of the olfactory bulb   1–9, 101–113

IPR   interpeduncular nucleus, rostral subnucleus   58–64, 101–102

IRt   intermediate reticular nucleus   76–100, 104–113

IS   inferior salivatory nucleus   79–82

IVF   interventricular foramen   107

K

KF   Kölliker-Fuse nucleus   71–75, 115–116

L

LA   lateroanterior hypothalamic nucleus   34–38, 103–106

La   lateral amygdaloid nucleus   37–40, 50, 128–131

LAcbSh   lateral accumbens shell   20–21, 23–25, 113–118

LaDL   lateral amygdaloid nucleus, dorsolateral part   41–51

Lat   lateral (dentate) cerebellar nucleus   78–84, 117–123

LatPC   lateral cerebellar nucleus, parvicellular part   81–83, 113–121

LaVL   lateral amygdaloid nucleus, ventrolateral part   43–49

LaVM   lateral amygdaloid nucleus, ventromedial part   46–51, 127

LC   locus coeruleus   75–79, 108–109

LD   laterodorsal thalamic nucleus   111

Ld   lambdoid septal zone   24–28, 101

LDDM   laterodorsal thalamic nucleus, dorsomedial part   39–45, 108–110

LDTg   laterodorsal tegmental nucleus   72–77, 102–107

LDTgV   laterodorsal tegmental nucleus, ventral part   71–73, 107

LDVL   laterodorsal thalamic nucleus, ventrolateral part   38–45, 112–118

LEnt   lateral entorhinal cortex   0, 49–73, 125–132

lfp   longitudinal fasciculus of the pons   66–72, 74, 103–109

LGP   lateral globus pallidus   31–45, 112–125

LH   lateral hypothalamic area   34–54, 107–113

LHb   lateral habenular nucleus   39–45, 48–49, 106

LHbL   lateral habenular nucleus, lateral part   46–47

LHbM   lateral habenular nucleus, medial part   46–47, 103–105

Li   linear nucleus of the medulla   88–90, 108

ll   lateral lemniscus   65–66, 68–73, 108–118

LM   lateral mammillary nucleus   53–56, 107–110

LMol   lacunosum moleculare layer of the hippocampus   42–64, 104–130

LO   lateral orbital cortex   0, 4–15, 17, 112–120

lo   lateral olfactory tract   1–30, 40, 109–124

LOT   nucleus of the lateral olfactory tract   33–38, 115–119

LPAG   lateral periaqueductal gray   59–74, 104–108

LPB   lateral parabrachial nucleus   72, 109–112

LPBC   lateral parabrachial nucleus, central part   73–77

LPBD   lateral parabrachial nucleus, dorsal part   74–76, 114

LPBE   lateral parabrachial nucleus, external part   73–75, 113–115
LPBI   lateral parabrachial nucleus, internal part   75
LPBS   lateral parabrachial nucleus, superior part   73
LPBV   lateral parabrachial nucleus, ventral part   73–75, 77–78
LPGi   lateral paragigantocellular nucleus   78–90, 104–113
LPLC   lateral posterior thalamic nucleus, laterocaudal part   54–55, 116–117
LPLR   lateral posterior thalamic nucleus, laterorostral part   46–53, 113–117
LPMC   lateral posterior thalamic nucleus, mediocaudal part   51–56, 115
LPMR   lateral posterior thalamic nucleus, mediorostral part   43–53, 107–114
LPO   lateral preoptic area   26–33, 105–109
LPtA   lateral parietal association cortex   43–48, 110–113
LR4V   lateral recess of the 4th ventricle   78–90, 109–117
LRt   lateral reticular nucleus   90–100, 108–114
LRtPC   lateral reticular nucleus, parvicellular part   90–92, 97–98, 114–116
LS   lateral septal nucleus   106
LSD   lateral septal nucleus, dorsal part   19–35, 102–106
LSI   lateral septal nucleus, intermediate part   18–32, 101–105
LSO   lateral superior olive   73–77, 110–113
LSS   lateral stripe of the striatum   19–29, 114, 117–123
LSV   lateral septal nucleus, ventral part   21–32, 105
LT   lateral terminal nucleus of the accessory optic tract   52–55
LV   lateral ventricle   15–55, 101, 103, 106–132
LVe   lateral vestibular nucleus   78–83, 111–115
LVPO   lateroventral periolivary nucleus   70–78, 110–113

M

M1   primary motor cortex   0, 11–42, 107–123
M2   secondary motor cortex   0, 10–26, 28–42, 103–120
m5   motor root of the trigeminal nerve   66, 68–73, 114–117
MA3   medial accessory oculomotor nucleus   56–61, 102–103

maopt   medial accessory optic tract   36–38
MCLH   magnocellular nucleus of the lateral hypothalamus   44–47, 111–113
mcp   middle cerebellar peduncle   66–80, 112–120
MCPC   magnocellular nucleus of the posterior commissure   53–55, 104–106
MCPO   magnocellular preoptic nucleus   27–38, 112–119
MD   mediodorsal thalamic nucleus   36–39, 48, 104
MDC   mediodorsal thalamic nucleus, central part   40–47, 104–105
MdD   medullary reticular nucleus, dorsal part   92–100, 109–112
MDL   mediodorsal thalamic nucleus, lateral part   40–47, 105–108
MDM   mediodorsal thalamic nucleus, medial part   40–47, 102–103
MdV   medullary reticular nucleus, ventral part   92–100, 104–110
ME   median eminence   44–49, 101
Me   medial amygdaloid nucleus   50
Me5   mesencephalic trigeminal nucleus   65–66, 68–78, 107–110
me5   mesencephalic trigeminal tract   73–77, 109–111
MeA   medial amygdaloid nucleus, anterior part   40–41, 115
MeAD   medial amygdaloid nucleus, anterior dorsal   37–39, 116–118
MeAV   medial amygdaloid nucleus, anteroventral part   39, 116, 118
Med   medial (fastigial) cerebellar nucleus   80–88, 104–107, 110–111
MedDL   medial cerebellar nucleus, dorsolateral protuberance   84–88, 108–113
MEnt   medial entorhinal cortex   0, 64–66, 68–74, 122–132
MePD   medial amygdaloid nucleus, posterodorsal part   42–50, 116–120
MePV   medial amygdaloid nucleus, posteroventral part   42–49, 115–118
mfb   medial forebrain bundle   15–44, 109–111, 114–116
mfba   medial forebrain bundle, 'a' component   108
MG   medial geniculate nucleus   63–64, 121
MGD   medial geniculate nucleus, dorsal part   54–59, 117–120
MGM   medial geniculate nucleus, medial part   54–59, 116
MGP   medial globus pallidus (entopeduncular nucleus)   40–44, 114–119

MGV   medial geniculate nucleus, ventral part   54–61, 117–120
MHb   medial habenular nucleus   38–49, 102–104
Mi   mitral cell layer of the olfactory bulb   1–9, 101–113
MiA   mitral cell layer of the accessory olfactory bulb   3, 5, 108, 110
Min   minimus nucleus   61
MiTg   microcellular tegmental nucleus   65–66, 68–72, 111–118
ML   medial mammillary nucleus, lateral part   53–57, 103–106
ml   medial lemniscus   45–86, 101–113
mlf   medial longitudinal fasciculus   59–92, 96–104
MM   medial mammillary nucleus, medial part   53–59, 101–105
MMn   medial mammillary nucleus, median part   53–55, 101
MnA   median accessory nucleus of the medulla   100
MnPO   median preoptic nucleus   26–30, 101–102
MnR   median raphe nucleus   64–73, 101
MO   medial orbital cortex   0, 4–14, 101–105
Mo5   motor trigeminal nucleus   71–75, 109–113
Mol   molecular layer of the dentate gyrus   39–65, 104–123, 125–129
mp   mammillary peduncle   55–58
MPA   medial preoptic area   25–36, 102–107
MPB   medial parabrachial nucleus   72–78, 107–112
MPBE   medial parabrachial nucleus external part   75–76, 113
MPOC   medial preoptic nucleus, central part   30–32
MPOL   medial preoptic nucleus, lateral part   32–35
MPOM   medial preoptic nucleus, medial part   30–35, 102–104
MPT   medial pretectal nucleus   53–54, 103–104
MPtA   medial parietal association cortex   0, 43–48, 108–110
MRe   mammillary recess of the 3rd ventricle   53–54
MS   medial septal nucleus   21–29, 101–103
MSO   medial superior olive   75
MT   medial terminal nucleus of the accessory optic tract   56–59, 108–109
mt   mammillothalamic tract   39–52, 105–106
mtg   mammillotegmental tract   52–67, 103–104
MTu   medial tuberal nucleus   46–48, 105–110
MVe   medial vestibular nucleus   77, 87–90, 113–115
MVeMC   medial vestibular nucleus, magnocellular part   78–86, 103–113
MVePC   medial vestibular nucleus, parvicellular part   78–86, 103–112

MVPO   medioventral periolivary nucleus   70–78, 106–109
MZMG   marginal zone of the medial geniculate   55–62

N

ns   nigrostriatal bundle   40–54, 107, 109, 111, 113

O

O   nucleus O   76–77
ocb   olivocochlear bundle   76–78, 101, 116, 119
ON   olfactory nerve layer   1–3, 109–113
Op   optic nerve layer of the superior colliculus   56–71, 101, 103–114, 116
OPC   oval paracentral thalamic nucleus   45–49, 107–109
OPT   olivary pretectal nucleus   51–54, 105–111
opt   optic tract   34–55, 110–124
Or   oriens layer of the hippocampus   39–62, 106–110, 112–130, 132
OT   nucleus of the optic tract   53–56, 108–115
OV   olfactory ventricle (olfactory part of lateral ventricle)   12–14
ox   optic chiasm   31–33, 101–109

P

P5   peritrigeminal zone   70–75, 111–112
P7   perifacial zone   78–85, 107–114
Pa4   paratrochlear nucleus   68–70, 103–106
Pa5   paratrigeminal nucleus   90–93, 97, 117
Pa6   paraabducens nucleus   77–79, 104–105
PaAP   paraventricular hypothalamic nucleus, anterior parvicellular part   36, 101–103
PaDC   paraventricular hypothalamic nucleus, dorsal cap   37–39, 102–103
PAG   periaqueductal gray   52–58, 101
PaLM   paraventricular hypothalamic nucleus, lateral magnocellular part   36–39, 102–104
PaMM   paraventricular hypothalamic nucleus, medial magnocellular part   37–39, 102–103
PaMP   paraventricular hypothalamic nucleus, medial parvicellular part   37–41, 101
PaPo   paraventricular hypothalamic nucleus, posterior part   40–41, 102–106
PaR   pararubral nucleus   108–111
PaS   parasubiculum   65–73, 121–124

xxviii

PaV paraventricular hypothalamic nucleus, ventral part 36–37, 101–102

PBG parabigeminal nucleus 65–66, 68–69, 118

PBP parabrachial pigmented nucleus 55–63, 104, 108–112

PBW parabrachial nucleus, waist part 76

PC paracentral thalamic nucleus 35–48, 105–108

pc posterior commissure 50–58, 101–104

PC5 parvicellular motor trigeminal nucleus 69–73, 114–115

PCGS paracochlear glial substance 76, 82, 117

pcn precentral fissure 76–80, 101–109

PCom nucleus of the posterior commissure 53–55, 104–105

PCRt parvicellular reticular nucleus 86–91, 109–115

PCRtA parvicellular reticular nucleus, alpha part 75–85, 109–115

pcuf preculminate fissure 76–81, 101–113

PDTg posterodorsal tegmental nucleus 78–79, 101–102

Pe periventricular hypothalamic nucleus 29–47

PeF perifornical nucleus 45–48, 107–108

PF parafascicular thalamic nucleus 48–52, 103–108

PFl paraflocculus 74–87, 125–132

pfs parafloccular sulcus 75–87

PH posterior hypothalamic area 46–53, 101–106

Pi pineal gland 63

PIL posterior intralaminar thalamic nucleus 54–59, 114–118

Pir piriform cortex 0, 10–54, 106–132

PL paralemniscal nucleus 65–66, 68–71, 114

PLCo posterolateral cortical amygdaloid nucleus (C2) 41–52, 119–130

plf posterolateral fissure 75–82, 86–89, 101–110

PLi posterior limitans thalamic nucleus 54–58

PM paramedian lobule 85–98, 115–128

pm principal mammillary tract 53–57

PMCo posteromedial cortical amygdaloid nucleus (C3) 46–62, 116–129

PMD premammillary nucleus, dorsal part 51–53, 102–106

PMn paramedian reticular nucleus 90–96, 102–103

PMnR paramedian raphe nucleus 64–73, 102–104

PMV premammillary nucleus, ventral part 50–52, 103–107

PN paranigral nucleus 59–62, 103–104

Pn pontine nuclei 62–69, 101–111

PnC pontine reticular nucleus, caudal part 73–78, 101–108

PnO pontine reticular nucleus, oral part 64–72, 104–111

PnR pontine raphe nucleus 74–75

PnV pontine reticular nucleus, ventral part 73–75, 101–103

Po posterior thalamic nuclear group 42–53, 109–118

PoDG polymorph layer of the dentate gyrus 40–64, 104–106, 108, 110–112, 114–118, 120–126

PoMn posteromedian thalamic nucleus 47–48

Post postsubiculum 111–121

PoT posterior thalamic nuclear group, triangular part 54–59, 113–114

PP peripeduncular nucleus 53–61, 115–120

ppf prepyramidal fissure 84–98, 101–118, 120–124

PPT posterior pretectal nucleus 52–57, 105–114

PPTg pedunculopontine tegmental nucleus 65–66, 68–72, 109–114

PPy parapyramidal nucleus 78–85

PR prerubral field 50–54, 106

Pr prepositus nucleus 80–89, 101–102

Pr5 principal sensory trigeminal nucleus 71–72, 119

Pr5DM principal sensory trigeminal nucleus, dorsomedial part 73–78, 115–118

Pr5VL principal sensory trigeminal nucleus, ventrolateral part 73–78, 114–118

prb Probst's bundle 80

PrC precommissural nucleus 49–51, 102–105

prf primary fissure 75–87, 101–123

PRh perirhinal cortex 0, 42–71, 127–132

PrL prelimbic cortex 0, 5–18, 101–106

PrS presubiculum 59–69, 123–126

PS parastrial nucleus 28–30, 105–106

psf posterior superior fissure 76–89, 94, 102–125

PSol parasolitary nucleus 91–93, 106

PSTh parasubthalamic nucleus 48–52, 108–111

PT paratenial thalamic nucleus 34–38, 101–104

PtA parietal association cortex 114–117

PV paraventricular thalamic nucleus 39–48, 101–103

pv periventricular fiber system 49–50, 101

PVA paraventricular thalamic nucleus, anterior part 33–38, 101–103

PVG periventricular gray 102

PVP paraventricular thalamic nucleus, posterior part 49–50, 102

Py pyramidal cell layer of the hippocampus 39–62, 104, 106–110, 112–128, 130, 132

py pyramidal tract 75–97, 101–106

pyx pyramidal decussation 98–102

R

R red nucleus 65, 103, 107

Rad stratum radiatum of the hippocampus 40–62, 104–128, 130–132

RAmb retroambiguus nucleus 98–99

Rbd rhabdoid nucleus 68, 71, 101

RC raphe cap 68–73

RCh retrochiasmatic area 39

Re reuniens thalamic nucleus 35–46, 101–105

ReIC recess of the inferior colliculus 101–102

REth retroethmoid nucleus 53–55, 112–113

rf rhinal fissure 6–73, 101–104, 107–130

Rh rhomboid thalamic nucleus 39–46, 101–104

RI rostral interstitial nucleus of medial longitudinal fasciculus 51–54, 103–106

RLi rostral linear nucleus of the raphe 56–63, 101–102

RMC red nucleus, magnocellular part 59–64, 104–106, 108–110

RMg raphe magnus nucleus 71–85, 101–103

Ro nucleus of Roller 89–95, 102–103

ROb raphe obscurus nucleus 83–99

RPa raphe pallidus nucleus 75–96, 101

RPC red nucleus, parvicellular part 55–59, 104–106, 110

RPF retroparafascicular nucleus 52–55, 105–106

RPO rostral periolivary region 68–72, 106–114

RR retrorubral nucleus 66–67, 115–117

RRF retrorubral field 62–65, 109–111

RRF/A8 retrorubral fields/A8 dopamine cells 105–108, 112

rs rubrospinal tract 62–100, 112–114, 116

RSA retrosplenial agranular cortex 0, 39–72, 101–107, 109–126

RSG retrosplenial granular cortex 0, 39–66, 108–117

RSGa retrosplenial granular a cortex 103–107

RSGb retrosplenial granular b cortex 101–104, 107

Rt reticular thalamic nucleus 35–49, 106–121

RtTg reticulotegmental nucleus of the pons 65–66, 68–75, 101–107

RtTgP reticulotegmental nucleus of the pons, pericentral part 66, 68–70, 102, 104–107

RVL rostroventrolateral reticular nucleus 84–91, 114–115

S

S subiculum 51–67, 104–132

S1 primary somatosensory cortex 14–20, 27, 48–50, 115–123, 125–132

S1BF primary somatosensory cortex, barrel field 0, 28–47, 118–132

S1DZ primary somatosensory cortex, dysgranular region 28–37, 39–42, 120, 122

S1FL primary somatosensory cortex, forelimb region 0, 23–34, 36–37, 116–122

S1HL primary somatosensory cortex, hindlimb region 0, 26–41, 111–117

S1J primary somatosensory cortex, jaw region 0, 16–23, 121–130

S1Sh primary somatosensory cortex, shoulder region 40–41, 115–119

S1ShNc primary somatosensory cortex, shoulder/neck region 0, 35, 38–39, 115–119

S1Tr primary somatosensory cortex, trunk region 0, 42–46, 110–119

S1ULp primary somatosensory cortex, upper lip region 0, 21–26, 121–131

S2 secondary somatosensory cortex 0, 23–46, 124, 129–132

s5 sensory root of the trigeminal nerve 71–75

Sag sagulum nucleus 70–73, 115

SC superior colliculus 55, 101–102, 117

Sc scaphoid thalamic nucleus 52

scc splenium of the corpus callosum 52, 102–104, 106

SCh suprachiasmatic nucleus 33–34, 37–38, 102

SChDM suprachiasmatic nucleus, dorsomedial part 35–36

SChVL suprachiasmatic nucleus, ventrolateral part 35–36

SCO subcommissural organ 50, 52–54, 101–102

SCom subcommissural nucleus 54, 105

scp superior cerebellar peduncle (brachium conjunctivum) 47–65, 71–79, 102–112

sf secondary fissure 91–110

SFi septofimbrial nucleus 29–36, 104

SFO subfornical organ 32–38, 101–102

SG suprageniculate thalamic nucleus 54–59, 116–117

SGl superficial glial zone of the cochlear nuclei 75–77, 79–85

SHi septohippocampal nucleus 15–26, 101–102

SHy septohypothalamic nucleus 27, 104–105

SI substantia innominata 27–38, 41–42, 109–119

SID substantia innominata, dorsal part 117

Sim simple lobule 73–84, 115–126

SL semilunar nucleus 15–18

SLEA sublenticular extended amygdala 31–33

SLEAC   sublenticular extended amygdala, central part   34–39
SLEAM   sublenticular extended amygdala, medial part   34–38
SLu   stratum lucidum, hippocampus   40–59, 106–113, 116–124
SM   nucleus of the stria medullaris   35, 110
sm   stria medullaris of the thalamus   33–43, 104–108
SMT   submammillothalamic nucleus   49–52
SMV   superior medullary velum   76–80, 101–102, 106
SNC   substantia nigra, compact part   52–63, 106–116
SNL   substantia nigra, lateral part   54–62, 116–118
SNR   substantia nigra, reticular part   51–64, 108–118
SO   supraoptic nucleus   36–39, 108–114
Sol   nucleus of the solitary tract   108–111, 113
sol   solitary tract   83–100, 104–108
SolC   nucleus of the solitary tract, commissural part   93–103
SolCe   nucleus of the solitary tract, central part   89–93
SolDL   solitary nucleus, dorsolateral part   90–97
SolDM   nucleus of the solitary tract, dorsomedial part   85–89
SolG   nucleus of the solitary tract, gelatinous part   90–95, 102
SolI   nucleus of the solitary tract, interstitial part   83–87, 89–91, 94–96
SolIM   nucleus of the solitary tract, intermediate part   83–95, 104–107
SolM   nucleus of the solitary tract, medial part   85–100, 102–107
SolV   solitary nucleus, ventral part   86–92, 94–99, 105–106
SolVL   nucleus of the solitary tract, ventrolateral part   86–99, 107
SOR   supraoptic nucleus, retrochiasmatic part   42–43
sox   supraoptic decussation   36–50, 101–121
sp5   spinal trigeminal tract   76–100, 108–120
Sp5C   spinal trigeminal nucleus, caudal part   94–100, 108–118
Sp5I   spinal trigeminal nucleus, interpolar part   85–95, 110–119
Sp5O   spinal trigeminal nucleus, oral part   78, 84, 116–118
Sp5ODM   spinal trigeminal nucleus, oral part, dorsomedial division   79–81
Sp5OVL   spinal trigeminal nucleus, oral part, ventrolateral division   79–83

SPa   subparaventricular zone of the hypothalamus   37–39
SPF   subparafascicular thalamic nucleus   48–50, 103–105
SPFPC   subparafascicular thalamic nucleus, parvicellular part   51–55, 106–112
Sph   sphenoid nucleus   76–77, 101–102
SPO   superior paraolivary nucleus   73–77, 106–109
SPTg   subpedencular tegmental nucleus   69–71, 104–108
SpVe   spinal vestibular nucleus   82–90, 107–115
st   stria terminalis   31–49, 108–125
StA   strial part of the preoptic area   30, 104–106
Stg   stigmoid hypothalamic nucleus   41
STh   subthalamic nucleus   45–50, 112–117
StHy   striohypothalamic nucleus   105–106
str   superior thalamic radiation   51–53, 116–120
Su3   supraoculomotor periaqueductal gray   62–66, 103–104
Su3C   supraoculomotor cap   61–66, 104–105
Su5   supratrigeminal nucleus   72–75, 110–114
Sub   submedius thalamic nucleus   39–46, 103–104
SubB   subbrachial nucleus   60–64, 116–119
SubCD   subcoeruleus nucleus, dorsal part   72–75, 109–112
SubCV   subcoeruleus nucleus, ventral part   72–75, 109–112
SubD   submedius thalamic nucleus, dorsal part   108
SubG   subgeniculate nucleus   50–55, 118–122
SubI   subincertal nucleus   42–45, 109–111
SubV   submedius thalamic nucleus, ventral part   108
SuG   superficial gray layer of the superior colliculus   56–71, 101–113
SuM   supramammillary nucleus   55–57, 103–106
SuML   supramammillary nucleus, lateral part   53–54, 107–109
SuMM   supramammillary nucleus, medial part   51–54, 101–102
sumx   supramammillary decussation   53–55, 101–102
SuVe   superior vestibular nucleus   78–81, 114–115

T

TC   tuber cinereum area   40–41, 102–103
Te   terete hypothalamic nucleus   46–50, 110
TeA   temporal association cortex   0, 46–71, 129–132
tfp   transverse fibers of the pons   66, 68–69, 101–111

TS   triangular septal nucleus   32–37, 102–104
ts   tectospinal tract   64–73, 96–100, 105
Tu   olfactory tubercle   13–31, 103–121
Tz   nucleus of the trapezoid body   71–78, 102–108
tz   trapezoid body   75–83, 101–107, 111–120

U

unc   uncinate fasciculus   76–78

V

V1   primary visual cortex   0, 49–74, 114–129
V2L   secondary visual cortex, lateral area   0, 47–72, 118–132
V2ML   secondary visual cortex, mediolateral area   0, 50–65, 108–117
V2MM   secondary visual cortex, mediomedial area   0, 49–72, 106–117
VA   ventral anterior thalamic nucleus   36–39, 108–110
VCA   ventral cochlear nucleus, anterior part   72–77, 119–121, 123
VCP   ventral cochlear nucleus, posterior part   78–84, 119–121
VDB   nucleus of the vertical limb of the diagonal band   19–25, 101–103
VeCb   vestibulocerebellar nucleus   80–83, 108–111
VEn   ventral endopiriform nucleus   31–51, 120–132
vert   vertebral artery   96–98
vhc   ventral hippocampal commissure   33–38, 101–105
VL   ventrolateral thalamic nucleus   39–45, 105–113
VLG   ventral lateral geniculate nucleus   45–47, 120–123
VLGMC   ventral lateral geniculate nucleus, magnocellular part   48–55
VLGPC   ventral lateral geniculate nucleus, parvicellular part   48–55
VLH   ventrolateral hypothalamic nucleus   108
VLL   ventral nucleus of the lateral lemniscus   65–66, 68, 112–118
VLPAG   ventrolateral periaqueductal gray   65–74, 104–107
VLPO   ventrolateral preoptic nucleus   30–32, 105–107
VLTg   ventrolateral tegmental area   68–69, 108–109

VM   ventromedial thalamic nucleus   39–49, 104–110
VMH   ventromedial hypothalamic nucleus   40–41, 48, 103–105, 107
VMHC   ventromedial hypothalamic nucleus, central part   42–47, 106
VMHDM   ventromedial hypothalamic nucleus, dorsomedial part   42–47, 102
VMHVL   ventromedial hypothalamic nucleus, ventrolateral part   42–47, 106, 108
VMPO   ventromedial preoptic nucleus   27–32, 102
vn   vomeronasal nerve   3
VO   ventral orbital cortex   0, 4–14, 105–111
VOLT   vascular organ of the lamina terminalis   26–28, 101
VP   ventral pallidum   15–34, 104–118
VPL   ventral posterolateral thalamic nucleus   38–50, 111–120
VPM   ventral posteromedial thalamic nucleus   40–52, 106–118
VRe   ventral reuniens thalamic nucleus   40–46, 103–104
vsc   ventral spinocerebellar tract   68–100, 109–117
VTA   ventral tegmental area   55–63, 103–109
VTg   ventral tegmental nucleus   70–73, 102–103
vtgx   ventral tegmental decussation   58–63, 101–103
VTM   ventral tuberomammillary nucleus   49–55, 110
VTRZ   visual tegmental relay zone   56–58
VTT   ventral tenia tecta   9–14, 102–106

X

X   nucleus X   82–88, 113–114, 116–117
Xi   xiphoid thalamic nucleus   36–45
xscp   decussation of the superior cerebellar peduncle   66–70, 101

Y

Y   nucleus Y   81, 115–118

Z

ZI   zona incerta   38–42, 55–56, 105–111, 120
ZID   zona incerta, dorsal part   43–54, 112–116
ZIV   zona incerta, ventral part   43–54, 112–119
ZL   zona limitans   25–29
Zo   zonal layer of the superior colliculus   56–71, 101–111, 114

Figure 0

AuD	2nd auditory cx, dorsal	DP	dorsal peduncular cx	M1	primary motor cx	PRh	perirhinal cx	S1HL	S1 cx, hindlimb region	TeA	temporal association cx
AuV	2nd auditory cx, ventral	Ect	ectorhinal cx	M2	2nd motor cx	PrL	prelimbic cx	S1J	S1 cx, jaw region	V1	primary visual cx
AIP	agran. insular cx, posterior	FrA	frontal association cx	MEnt	medial entorhinal cx	RSA	retrosplenial agran. cx	S1ShNc	S1 cx, shoulder/neck	V2L	2nd visual cx, lateral area
AID	agran. insular cx, dorsal	Cg1	cingulate cx, area 1	MO	medial orbital cx	RSG	retrosplenial granular	S1Tr	S1 cx, trunk region	V2ML	2nd visual cx, mediolat
AIV	agran. insular cx, ventral	Cg2	cingulate cx, area 2	MPtA	medial parietal ass cx	S1BF	S1 cx, barrel field	S1ULp	S1 cx, upper lip region	V2MM	2nd vis cx, mediomed
Au1	primary auditory cx	DLO	dorsolateral orbital cx	LEnt	lateral entorhinal cx	Pir	piriform cx	S2	2nd somatosensory cx	VO	ventral orbital cx
					LO	lateral orbital cx				

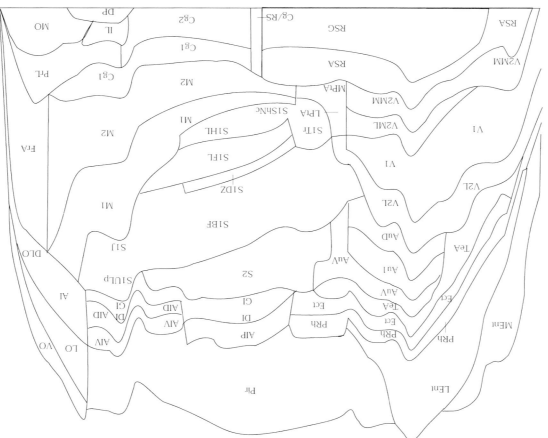

THE CORTEX OF THE MOUSE FLATTENED. The bottom of the map (x axis) represents the mediovental limit of the cortex at each coronal section of the atlas. The vertical axis (y axis) of the map represents distance from the mediovental margin measured around the surface of the cortex.

The map was constructed from the coronal sections by measuring the circumference, from the mediovental limit of the cortex (e.g. the corpus callosum) to each boundary between cortical regions. These data were collected for each coronal section and plotted against the anterior posterior position of the sections. The resulting graph was used as a template to draw the map. Minor irregularities were removed by drawing fair curves through the plotted points. Where the margin of the cortex is at the corpus callosum, the mediovental margin of the cortex approximates a straight line, and the distortion in the representation is minimal. However, at the rostral and caudal extremes of the brain the distortion in the representation increases because the mediovental margin of the cortex increasingly deviates from a straight line.

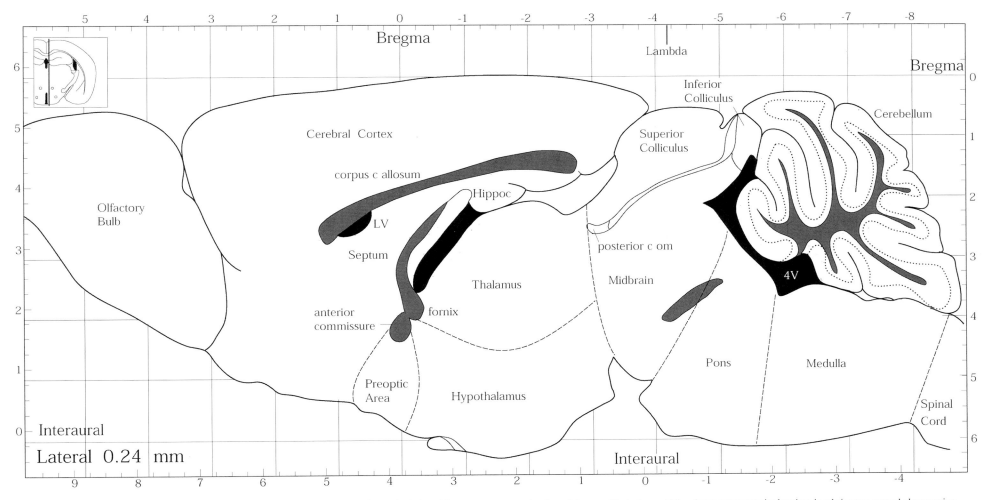

SAGITTAL SECTION OF THE MOUSE BRAIN   A drawing of a section of the mouse brain cut in a sagittal plane .24 mm lateral to the midline (Fig. 103 of this atlas). Ventricles are shown in black. The position of the skull marks (bregma, lambda and interaural) are indicated. The top grid line indicates anteroposterior locations in relation to the coronal plane passing through bregma. The bottom grid line shows anteroposterior locations in relation to a coronal plane passing through the interaural line. The left vertical grid line shows dorsoventral locations in relation to a horizontal plane passing through bregma. The right vertical grid shows dorsoventral positions in relation to a horizontal place passing through the interaural line.

xxxi

FIGURES

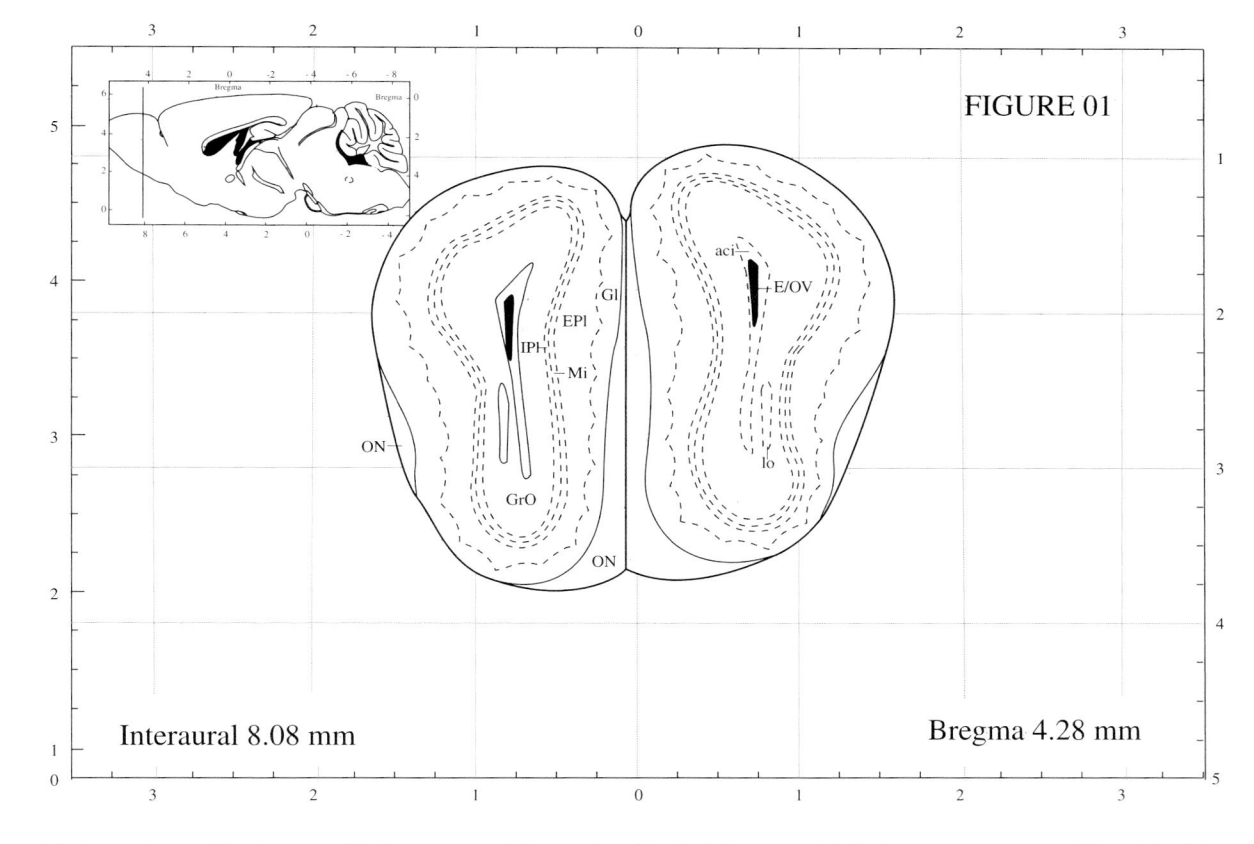

FIGURE 01

Interaural 8.08 mm

Bregma 4.28 mm

FIGURES 1 and 2
aci   ant comm, intrabulbar   EPl   ext plex layer olf bulb   GrA   granule cell lr acc olf b   IPl   int plex layer olf bulb

E/OV   ependy/olfactory ventr   Gl   glom layer olf bulb   GrO   granular cell lr olf   bulb   lo   lat olfactory tr   Mi   mitral cell layer olf bulb   ON olf actory n layer

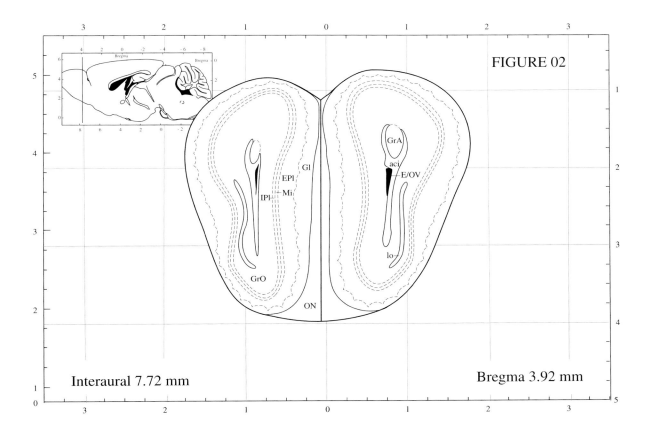

FIGURE 02

Interaural 7.72 mm   Bregma 3.92 mm

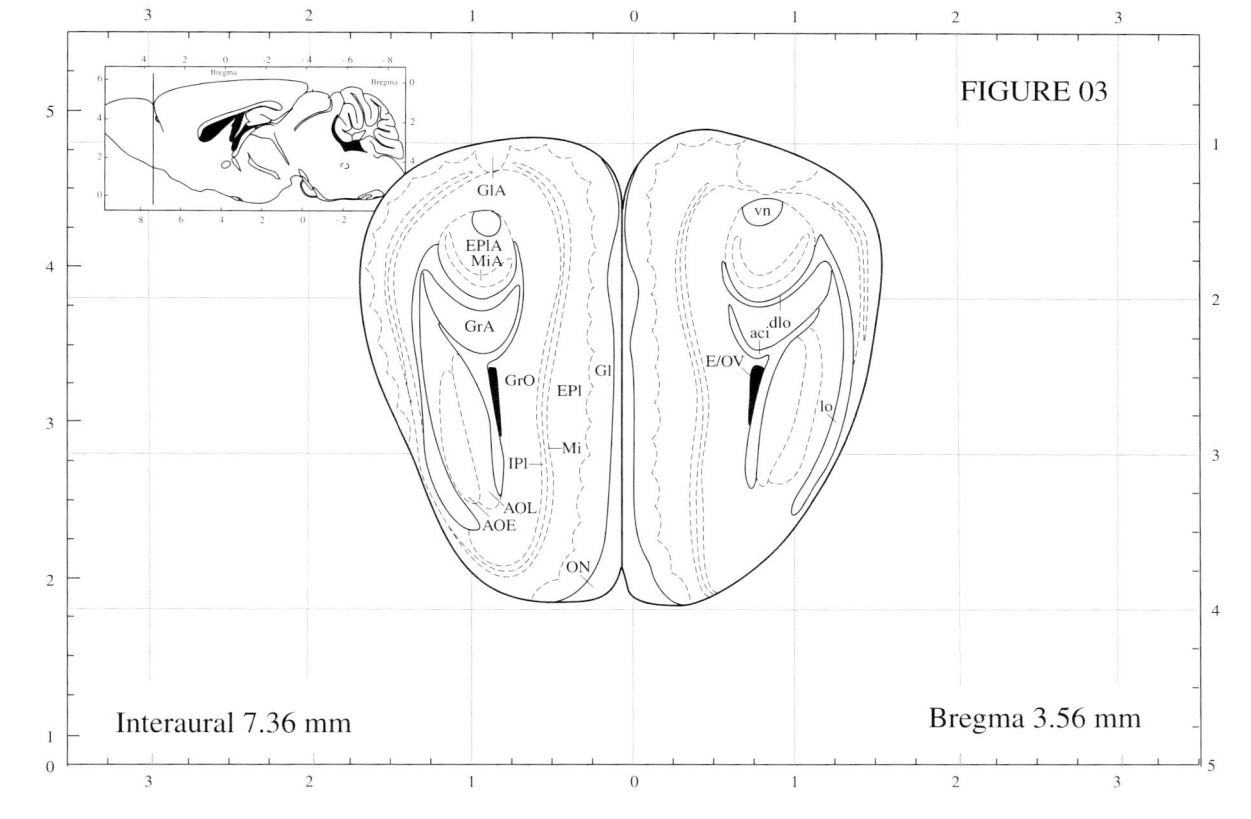

FIGURE 03

Interaural 7.36 mm

Bregma 3.56 mm

FIGURES 3 and 4
aci   ant comm, intrabulbar
AOB   acess olfactory bullb
AOD   ant olfactory nu, dors

AOE   ant olfactory nu, ext
AOL   ant olfactory nu, lat
AOM   ant olfactory nu, med

AOV   ant olfactory nu, vent
dlo   dors lat olfactory tr
E/OV   ependy/olfactory ventr

EPl   ext plex lr olf bulb
FrA   frontal assoc cortex
Gl   glom lr olf bulb

GrO   granular cell lr olf bulb
IPl   int plex lr olf bulb
LO   lat orbital cortex

lo   lat olfactory tr
Mi   mitral cell lr olf bulb
MO   med orbital cortex

ON olf actory n lr
VO   vent orbital cortex
vn   vomeronasal n

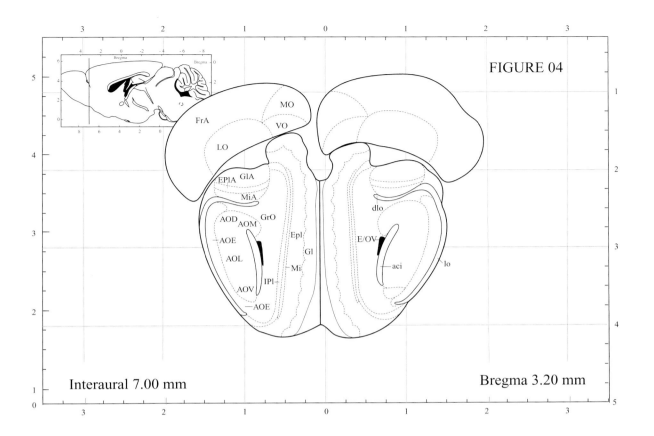

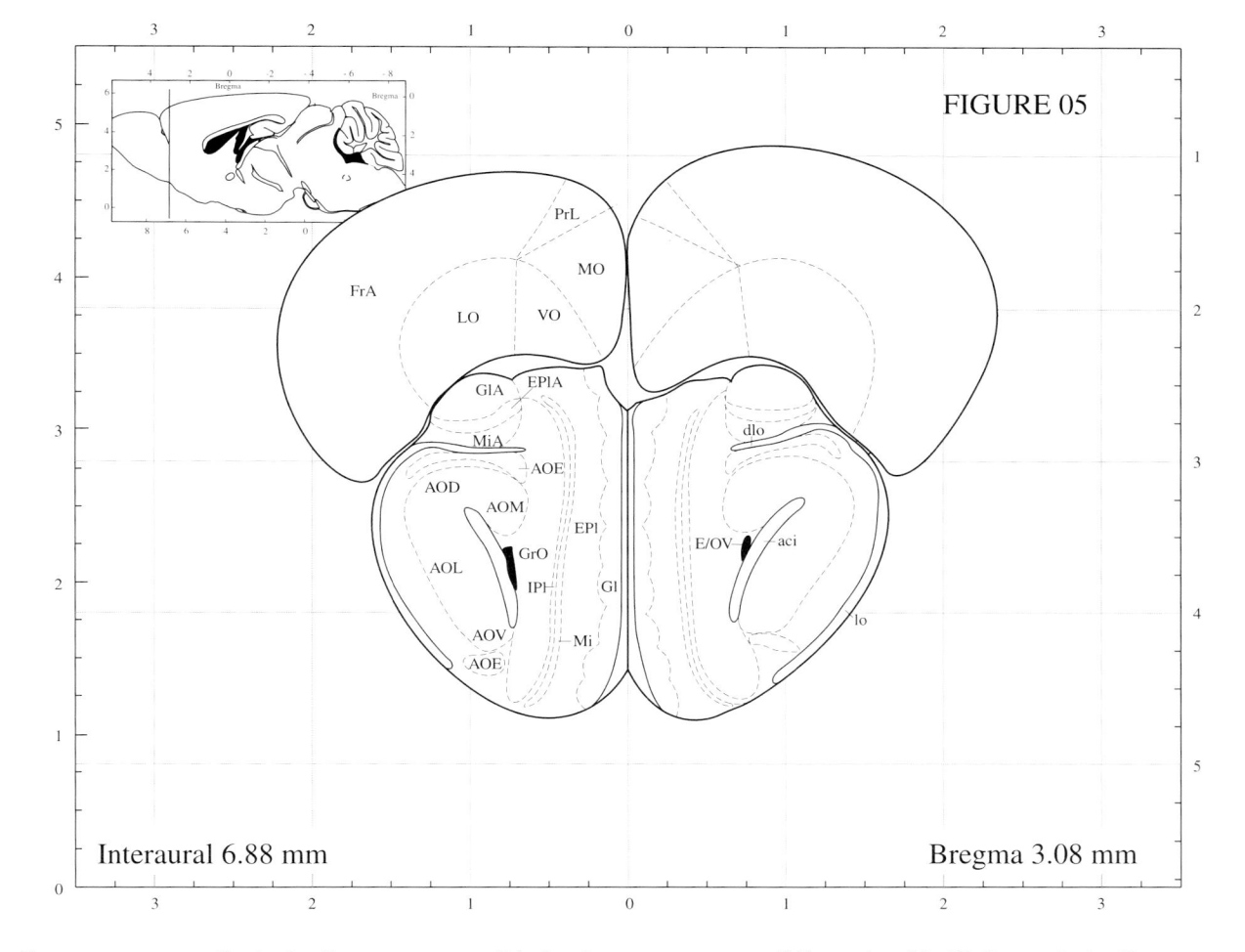

FIGURE 05

Interaural 6.88 mm

Bregma 3.08 mm

FIGURES 5 and 6
aci   ant comm, intrabulbar
AOB   acess olfactory bullb
AOD   ant olfactory nu, dors

AOE   ant olfactory nu, ext
AOL   ant olfactory nu, lat
AOM   ant olfactory nu, med
AOV   ant olfactory nu, vent

dlo dors lat olfactory tr
DLO   dorsolat orbital cortex
E/OV   ependy/olfactory ventr
EPl   ext plex lr olf bulb

FrA   frontal assoc cortex
Gl   glom lr olf bulb
GlA   glom lr acc olf bulb

GrO   granular cell lr olf bulb
IPl   int plex lr olf bulb
LO   lat orbital cortex

lo   lat olfactory tr
Mi   mitral cell lr olf bulb
MO   med orbital cortex

PrL   prelimbic cortex
rf   rhinal fissure
VO   vent orbital cortex

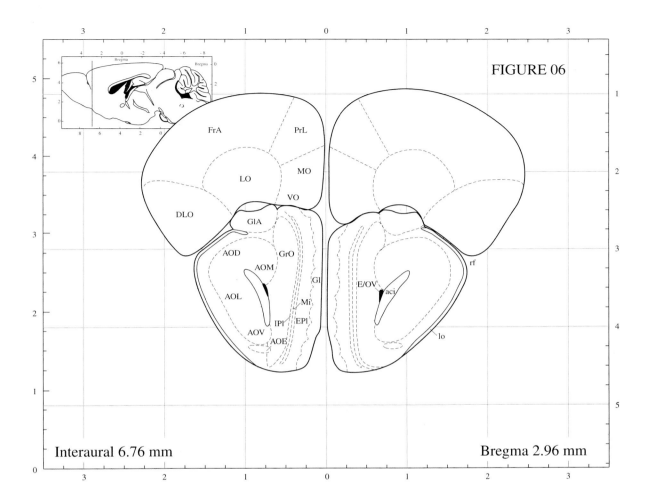

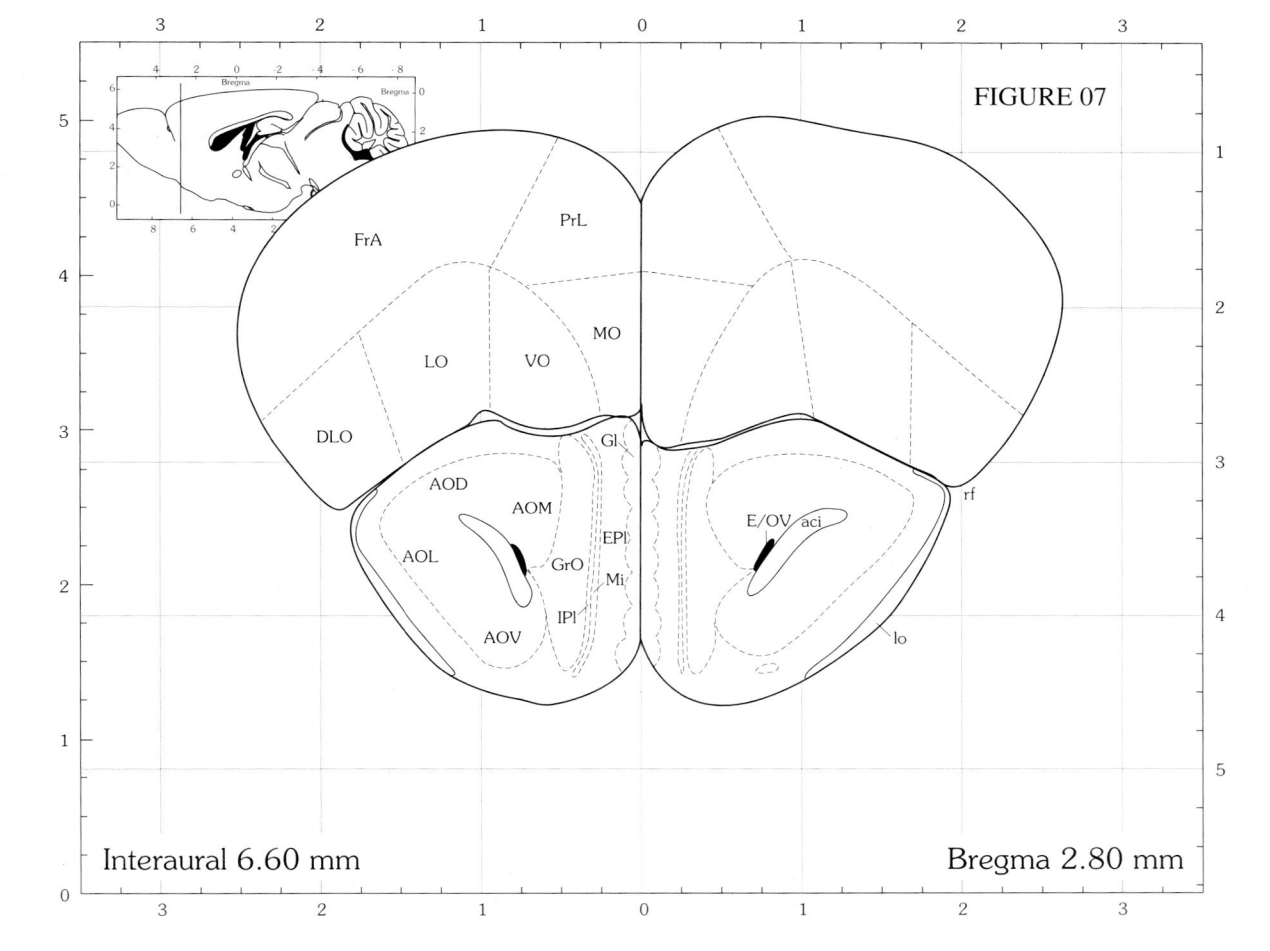

FIGURE 07

Interaural 6.60 mm

Bregma 2.80 mm

FIGURES 7 and 8
aci   ant comm, intrabulbar
AOD   ant olfactory nu, dors

AOL   ant olfactory nu, lat
AOM   ant olfactory nu, med
AOV   ant olfactory nu, vent

DLO   dorsolat orbital cortex
E/OV   ependy/olfactory ventr
EPl   ext plex lr olf bulb

FrA   frontal assoc cortex
Gl   glom lr olf bulb
GrO   granular cell lr olf bulb

IPl   int plex lr olf bulb
LO   lat orbital cortex
lo   lat olfactory tr

Mi   mitral cell lr olf bulb
MO   med orbital cortex
PrL   prelimbic cortex

rf   rhinal fissure
VO   vent orbital cortex

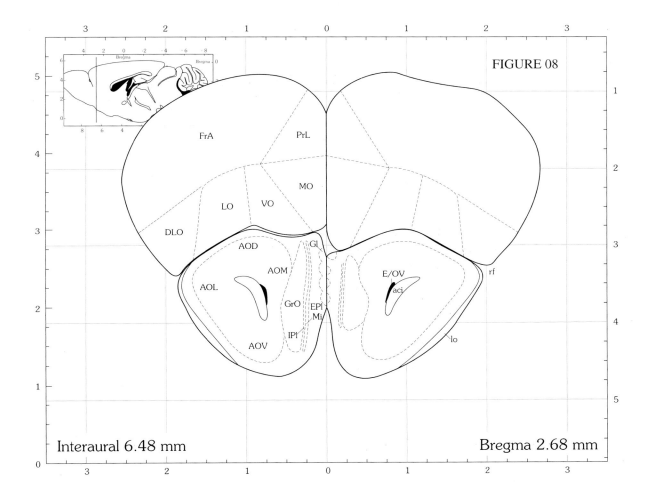

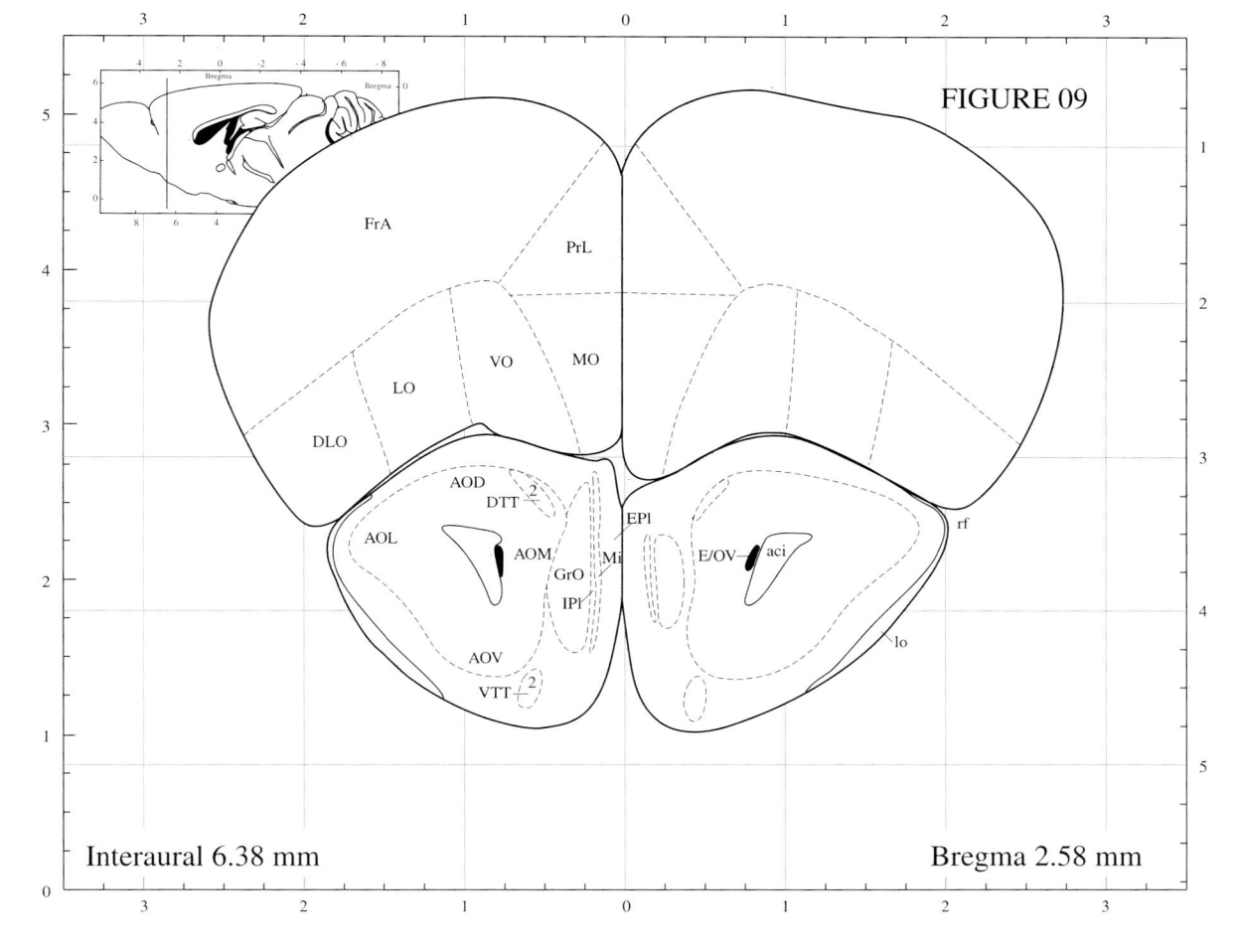

FIGURE 09

Interaural 6.38 mm

Bregma 2.58 mm

FIGURES 9 and 10
1, 2, 3, 4 layer 1, 2, 3, 4
aci ant comm, intrabulbar
AI agran insular cx

AOD ant olfactory nu, dors
AOL ant olfactory nu, lat
AOM ant olfactory nu, med
AOV ant olfactory nu, vent

DEn dors endopiriform nu
DLO dorsolat orbital cortex
DTT dors tenia tecta
E/OV ependy/olfactory ventr

EPl ext plex lr olf bulb
FrA frontal assoc cortex
GrO granular cell lr olf bulb
IPl int plex lr olf bulb

LO lat orbital cortex
lo lat olfactory tr
Mi mitral cell lr olf bulb
M2 2nd motor cortex

MO med orbital cortex
Pir piriform cortex
PrL prelimbic cortex

rf rhinal fissure
VO vent orbital cortex
VTT vent tenia tecta

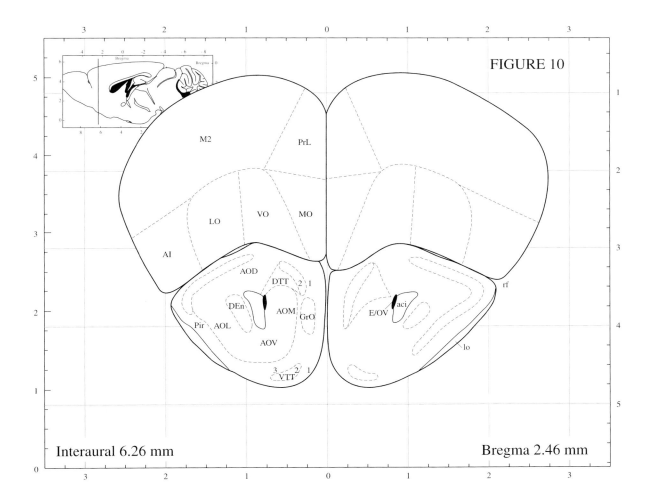

FIGURE 10

Interaural 6.26 mm   Bregma 2.46 mm

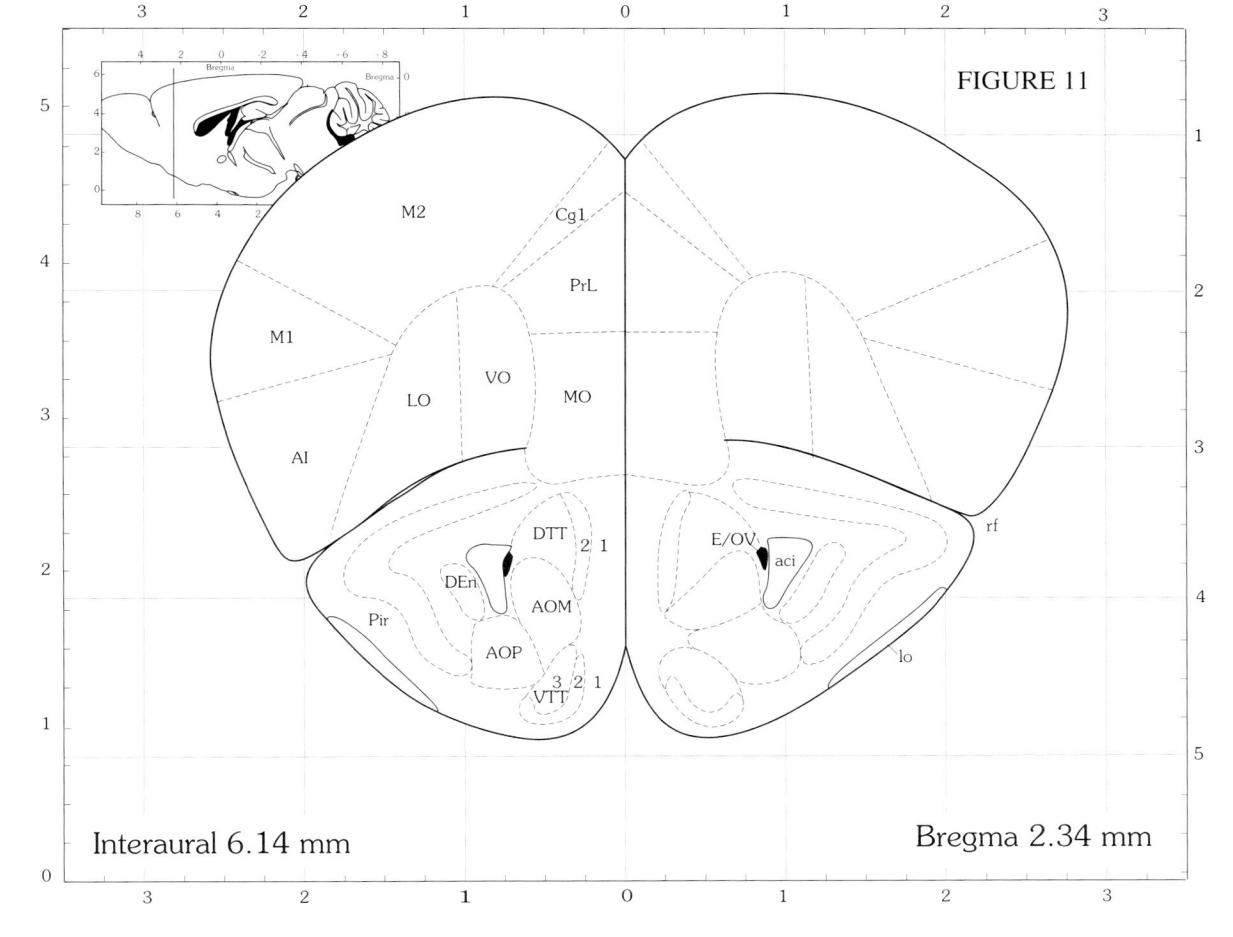

FIGURE 11

Interaural 6.14 mm

Bregma 2.34 mm

FIGURES 11 and 12
l, 2, 3, 4   layer 1, 2, 3, 4
aci   ant comm, intrabulbar

AI   agran insular cx
AOM   ant olfactory nu, med
AOP   ant olfactory nu, post

Cg1   cingulate cortex, area 1
DEn   dors endopiriform nu
DTT   dors tenia tecta

E/OV   ependy/olfactory ventr
LO   lat orbital cortex
lo   lat olfactory tr

M1   primary motor cortex
M2   2nd motor cortex
MO   med orbital cortex

OV olf actory ventric
Pir   piriform cortex
PrL   prelimbic cortex

rf   rhinal fissure
VO   vent orbital cortex
VTT   vent tenia tecta

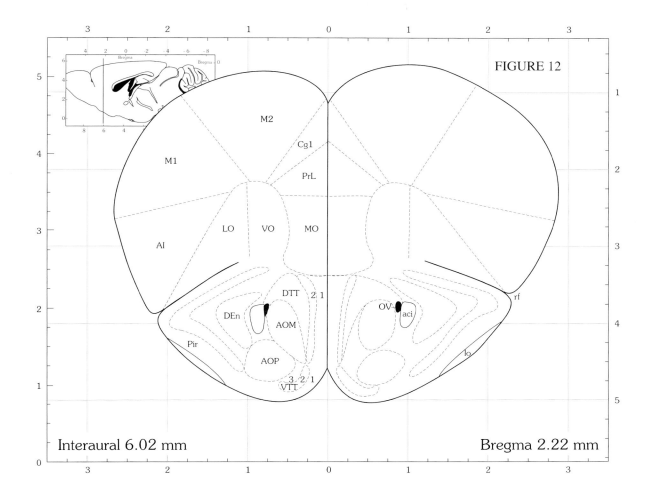

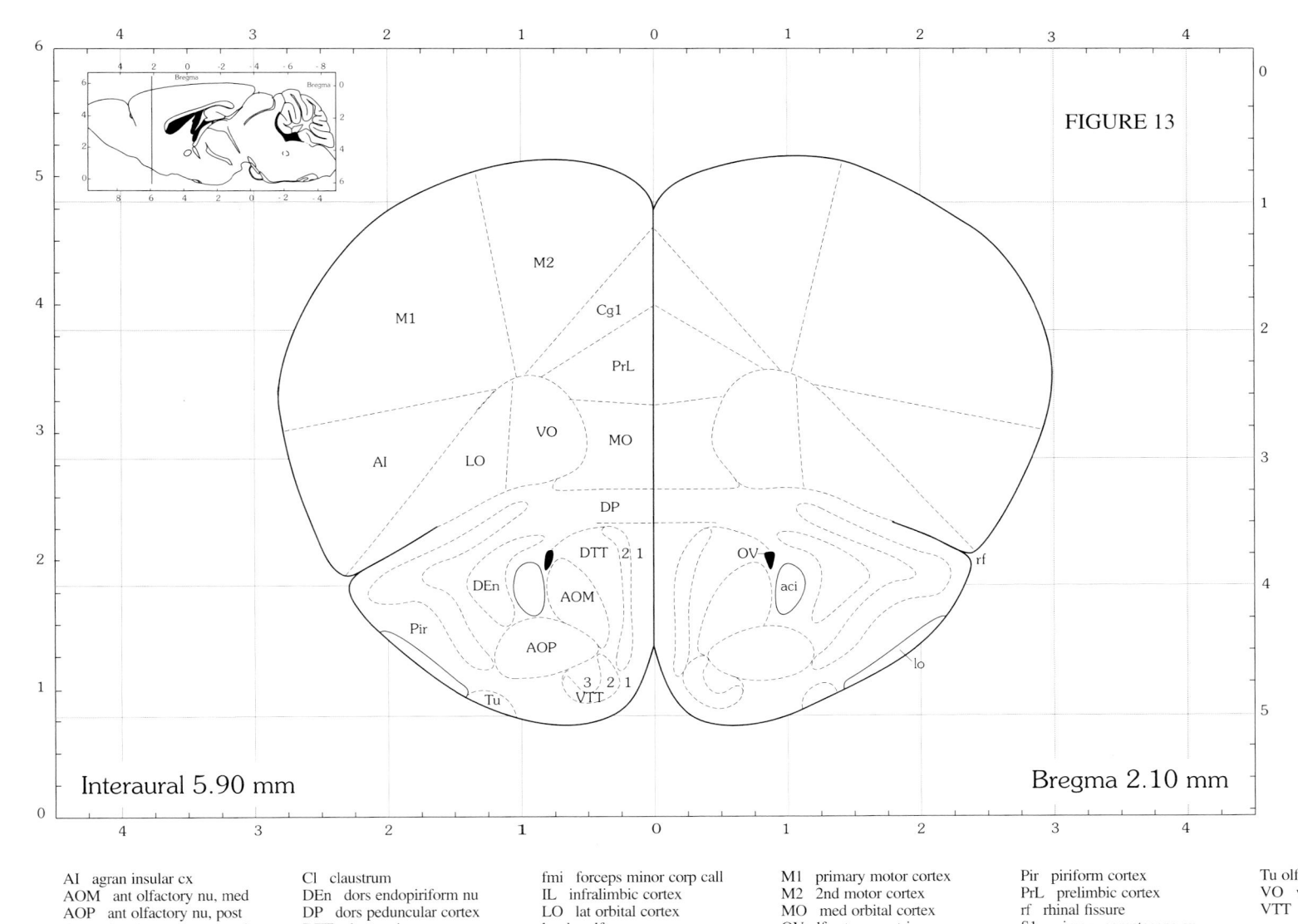

FIGURE 13

Interaural 5.90 mm

Bregma 2.10 mm

FIGURES 13 and 14
1, 2, 3, 4   layer 1, 2, 3, 4
aca   ant comm, ant
Acb   accumbens nu

AI   agran insular cx
AOM   ant olfactory nu, med
AOP   ant olfactory nu, post
Cg1   cingulate cortex, area 1

Cl   claustrum
DEn   dors endopiriform nu
DP   dors peduncular cortex
DTT   dors tenia tecta

fmi   forceps minor corp call
IL   infralimbic cortex
LO   lat orbital cortex
lo   lat olfactory tr

M1   primary motor cortex
M2   2nd motor cortex
MO   med orbital cortex
OV   olfactory ventric

Pir   piriform cortex
PrL   prelimbic cortex
rf   rhinal fissure
S1   primary somatosens cx

Tu olf actory tubercle
VO   vent orbital cortex
VTT   vent tenia tecta

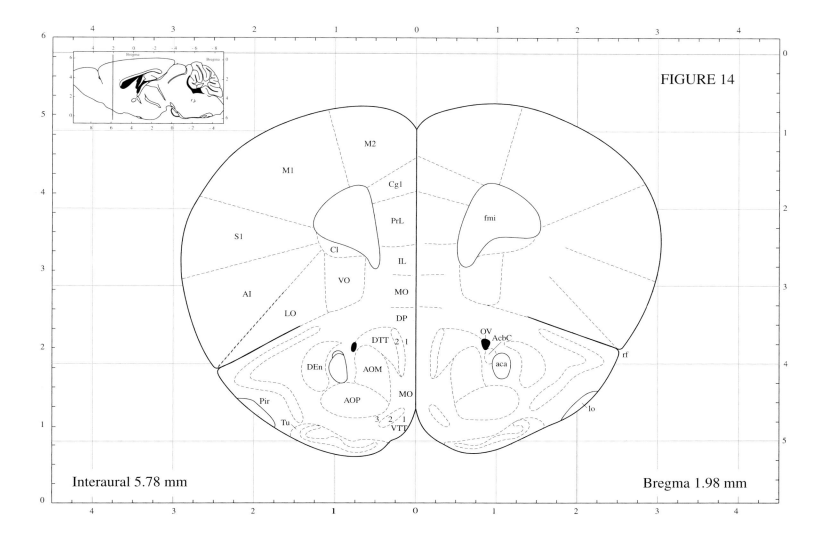

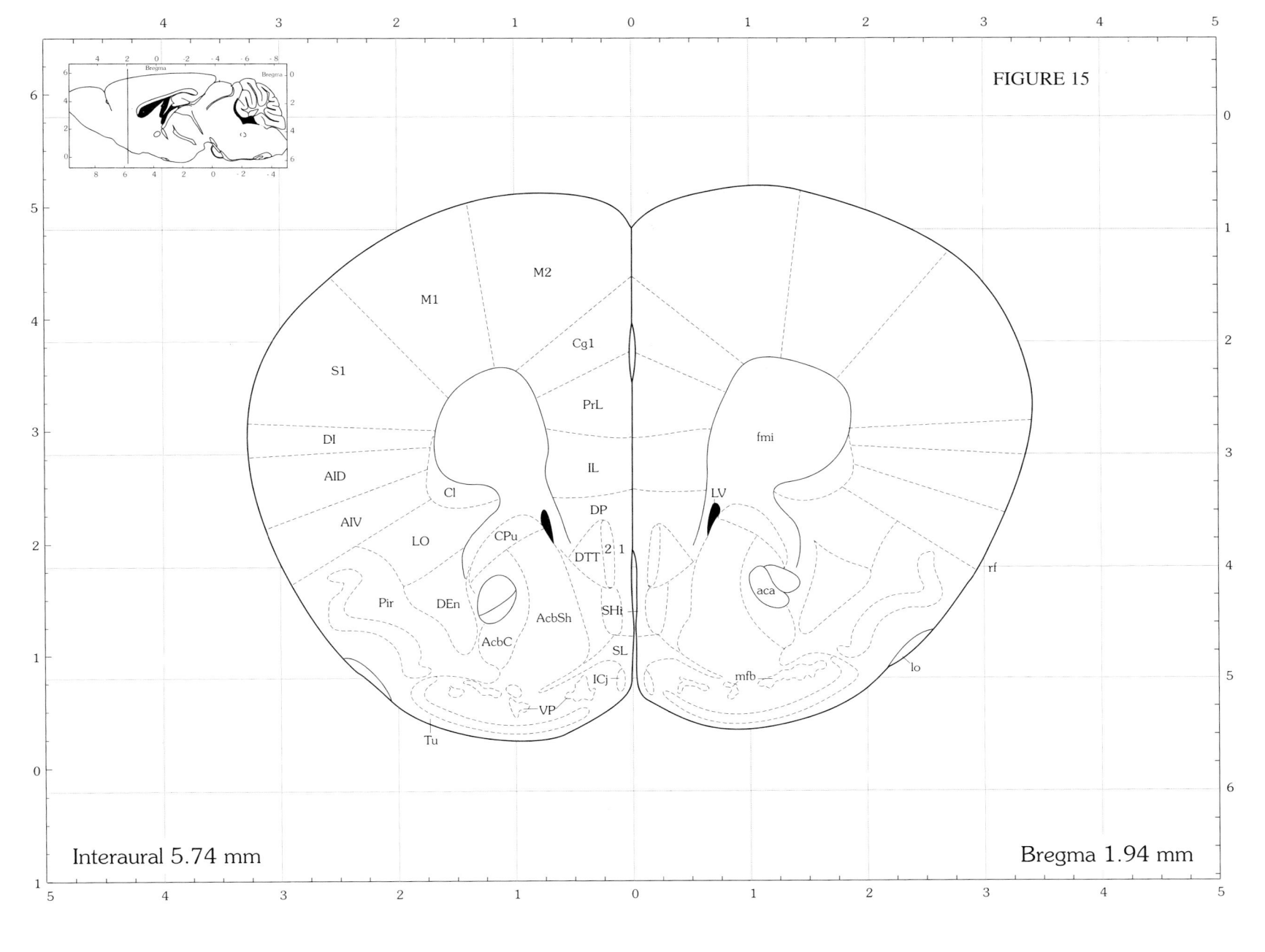

FIGURE 15

Interaural 5.74 mm

Bregma 1.94 mm

FIGURES 15 and 16
l, 2, 3, 4   layer 1, 2, 3, 4
aca   ant comm, ant
AcbC   accumbens nu, core
AcbSh   accumbens nu, shell

AID   agran insular cx dors
AIV   agran insular cx, vent
Cg1   cingulate cortex, area 1
Cl   claustrum
CPu   caudate putamen

DEn   dors endopiriform nu
DI   dysgran insular cx
DP   dors peduncular cortex
DTT   dors tenia tecta
fmi   forceps minor corp call

ICj   isl, Calleja
IL   infralimbic cortex
LO   lat orbital cortex
lo   lat olfactory tr
LV   lat ventric

M1   primary motor cortex
M2   2nd motor cortex
mfb   med forebrain bundle
Pir   piriform cortex

PrL   prelimbic cortex
rf   rhinal fissure
S1   primary somatosens cx
S1J   S1 cx, jaw reg

SHi   septohippocampal nu
SL   semilunar nu
Tu   olf actory tubercle
VP   vent pallidum

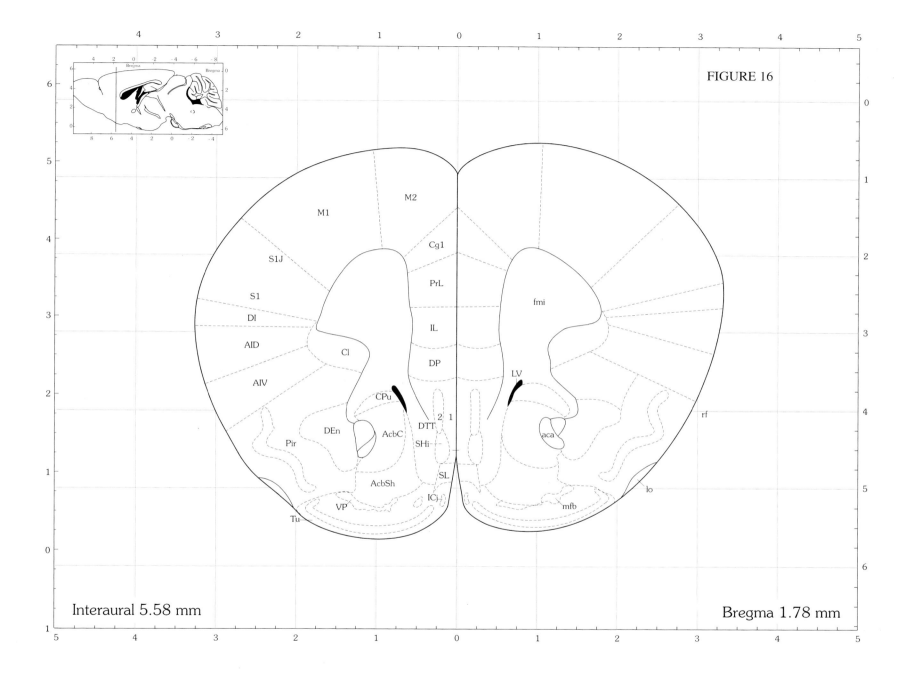

FIGURE 17

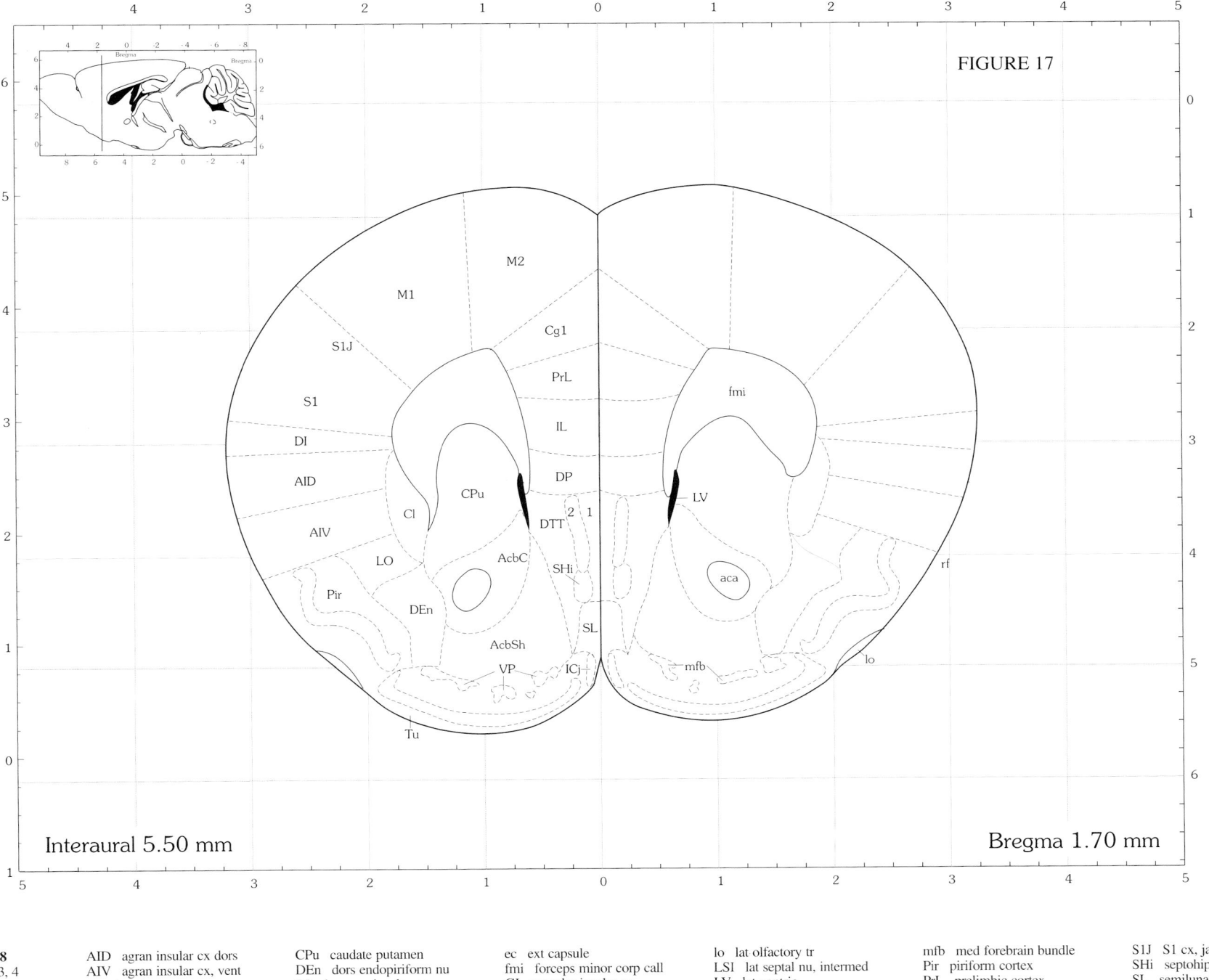

Interaural 5.50 mm

Bregma 1.70 mm

FIGURES 17 and 18
l, 2, 3, 4   layer 1, 2, 3, 4
aca   ant comm, ant
AcbC   accumbens nu, core
AcbSh   accumbens nu, shell

AID   agran insular cx dors
AIV   agran insular cx, vent
cg   cingulum
Cg1   cingulate cortex, area 1
Cl   claustrum

CPu   caudate putamen
DEn   dors endopiriform nu
DI   dysgrom insular cx
DP   dors peduncular cortex
DTT   dors tenia tecta

ec   ext capsule
fmi   forceps minor corp call
GI   granular insular cx
ICj   isl, Calleja
IL   infralimbic cortex

lo   lat olfactory tr
LSI   lat septal nu, intermed
LV   lat ventric
M1   primary motor cortex
M2   2nd motor cortex

mfb   med forebrain bundle
Pir   piriform cortex
PrL   prelimbic cortex
rf   rhinal fissure
S1   primary somatosens cx

S1J   S1 cx, jaw reg
SHi   septohippocampal nu
SL   semilunar nu
Tu   olf actory tubercle
VP   vent pallidum

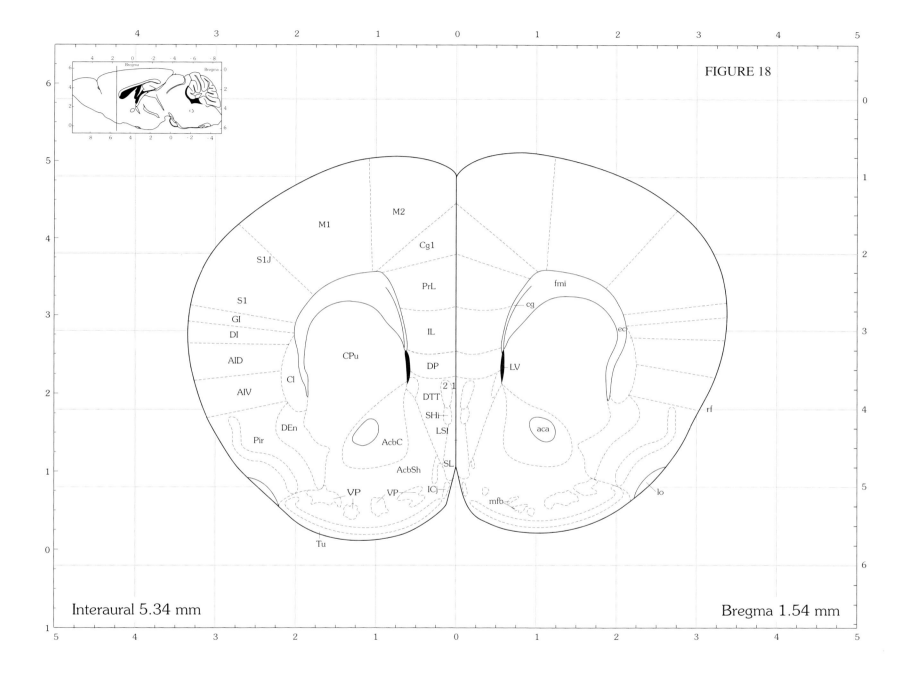

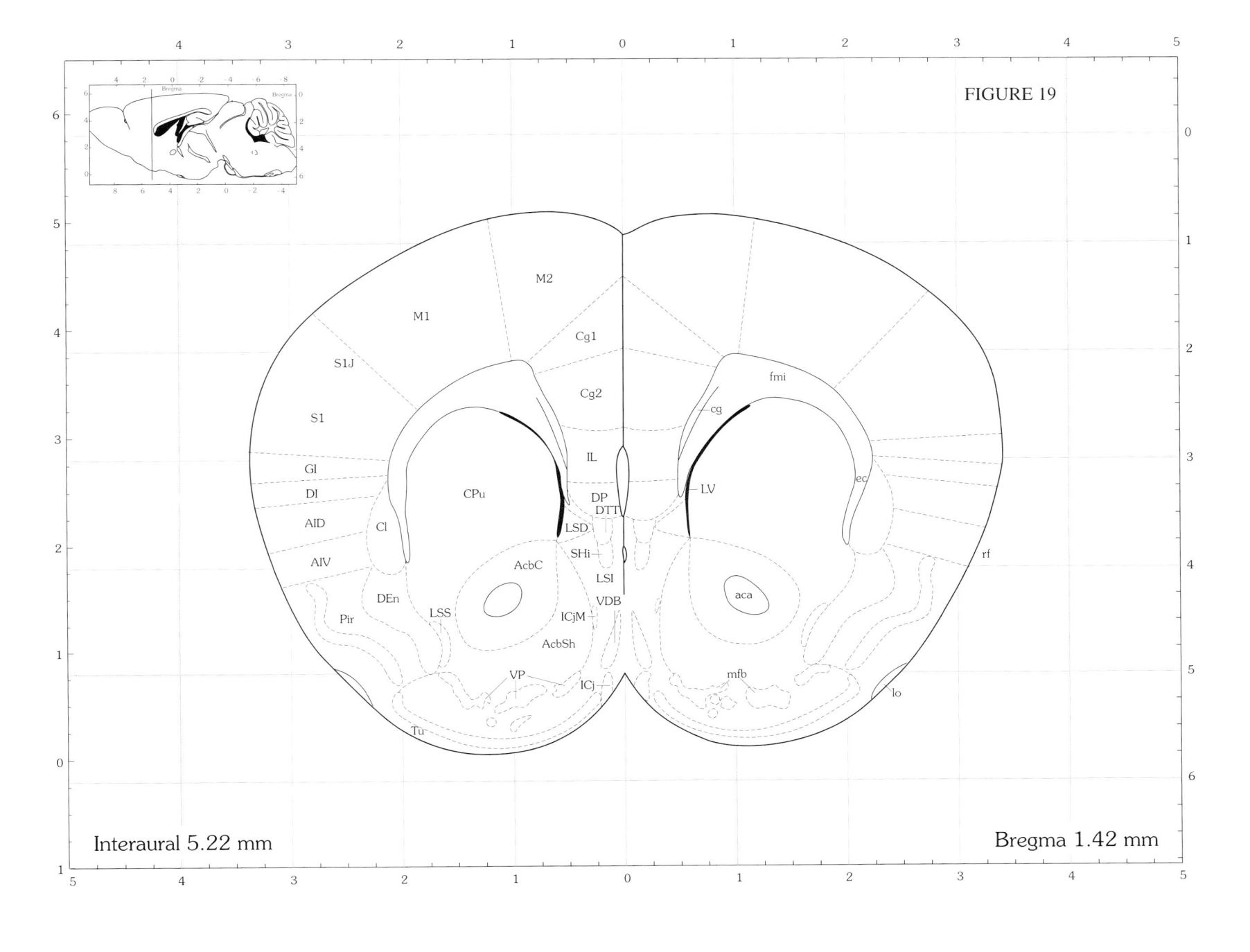

FIGURE 19

Interaural 5.22 mm

Bregma 1.42 mm

FIGURES 19 and 20
aca   ant comm, ant
AcbC   accumbens nu, core

AcbSh   accumbens nu, shell
AID   agran insular cx dors
AIV   agran insular cx, vent

cg   cingulum
Cg1   cingulate cortex, area 1
Cg2   cingulate cortex, area 2

Cl   claustrum
CPu   caudate putamen
DEn   dors endopiriform nu

DI   dysgran insular cx
DP   dors peduncular cortex
DTT   dors tenia tecta

ec   ext capsule
fmi   forceps minor corp call
GI   granular insular cx

ICj   isl, Calleja
ICjM   isl, Calleja, major isl
IL   infralimbic cortex

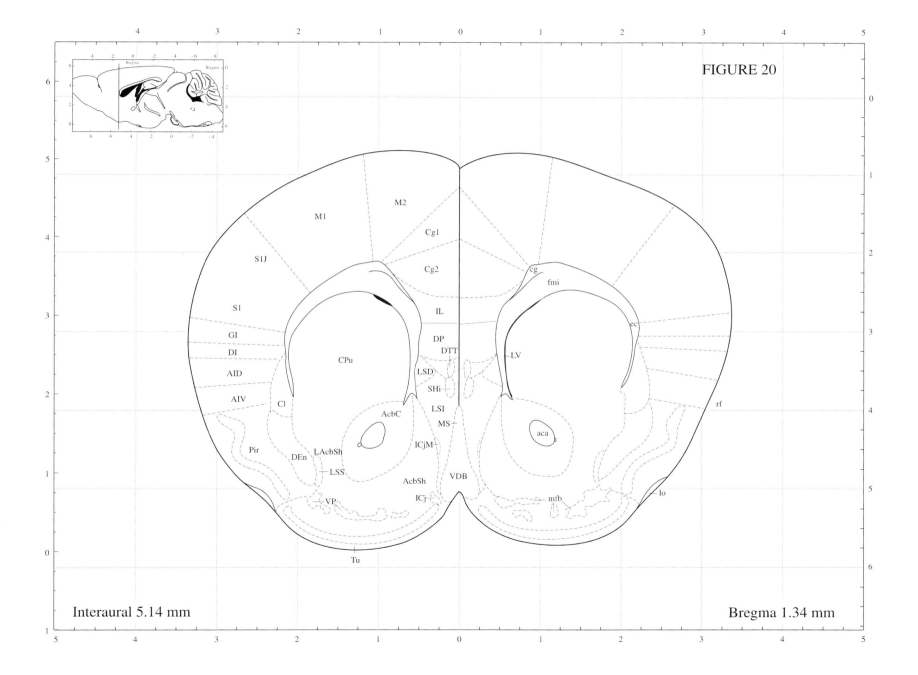

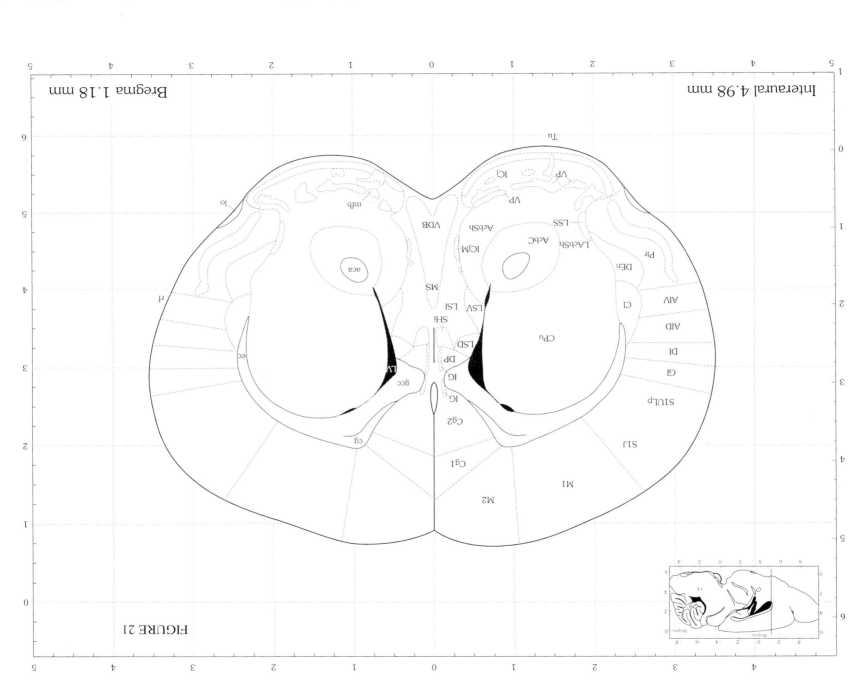

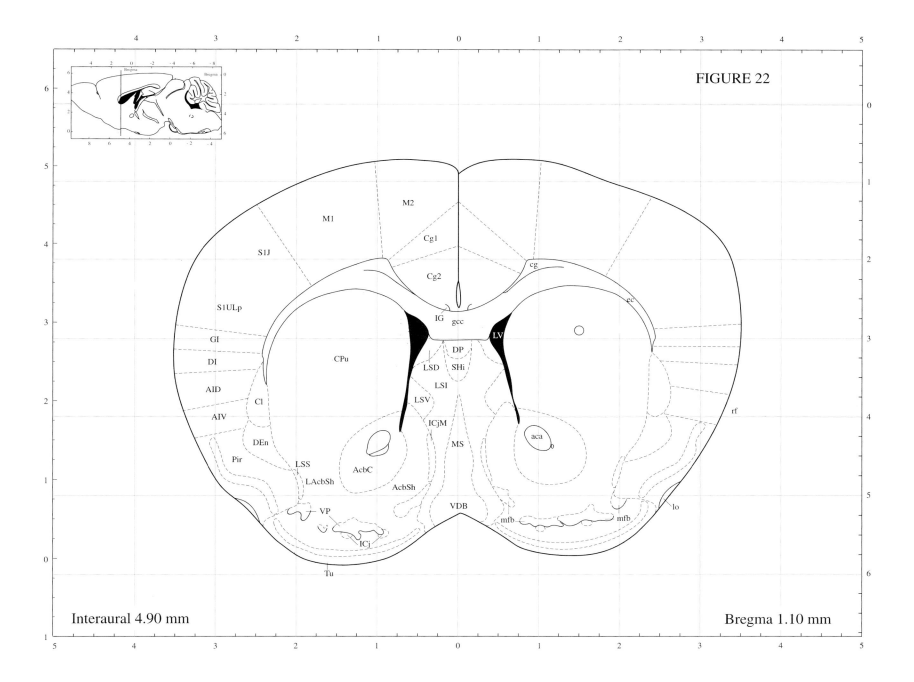

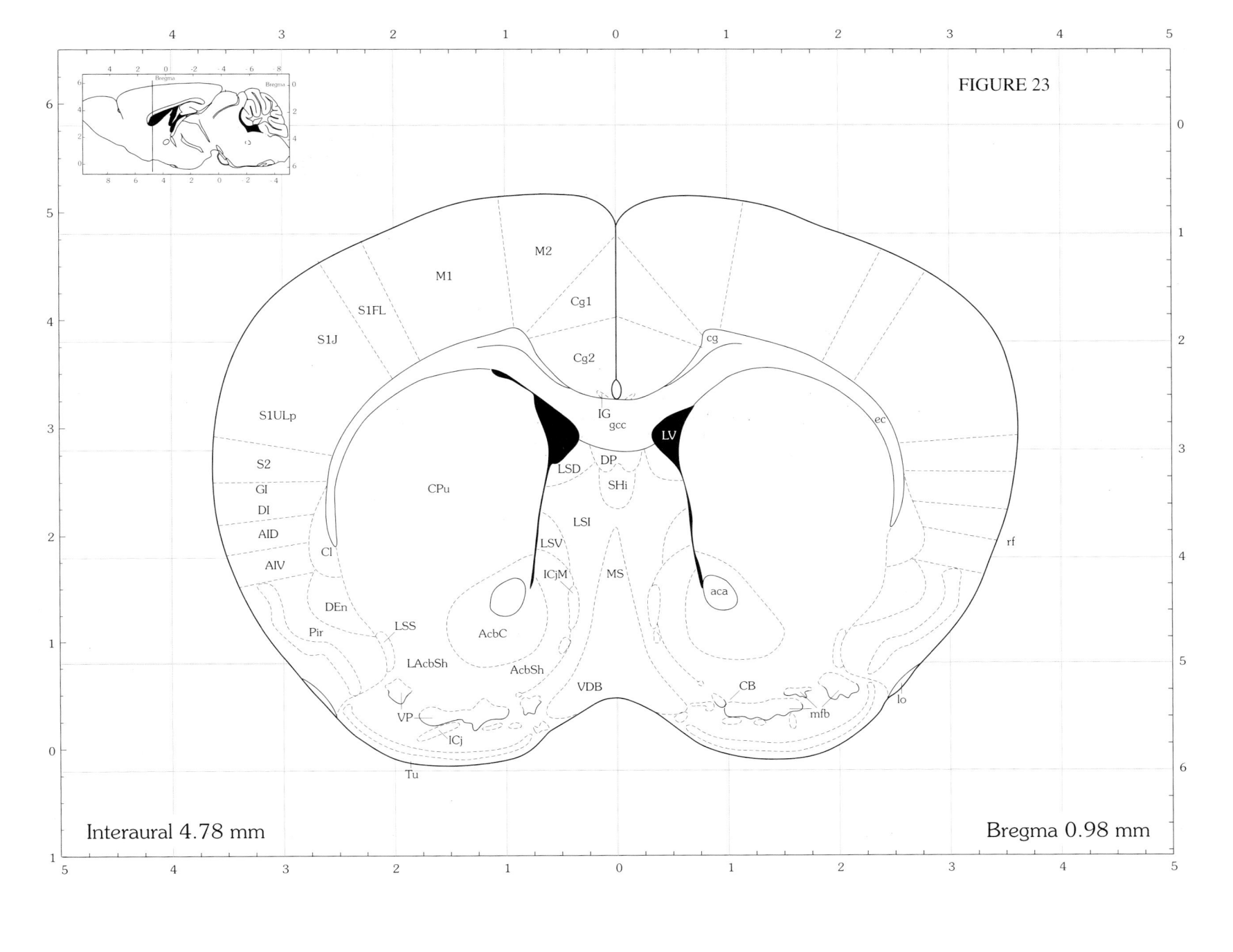

FIGURE 23

Interaural 4.78 mm

Bregma 0.98 mm

FIGURES 23 and 24
aca   ant comm, ant
AcbC   accumbens nu, core

AcbSh   accumbens nu, shell
AID   agran insular cx dors
AIV   agran insular cx, vent

CB   cell bridges, vent striat
cg   cingulum
Cg1   cingulate cortex, area 1

Cg2   cingulate cortex, area 2
Cl   claustrum
CPu   caudate putamen

DEn   dors endopiriform nu
DI   dysgran insular cxec ext
capsule

gcc   genu, corpus callosum
GI   granular insular cx
HDB   nu horiz limb diag bd

ICj   isl, Calleja
ICjM   isl, Calleja, major isl
IG   indusium griseum

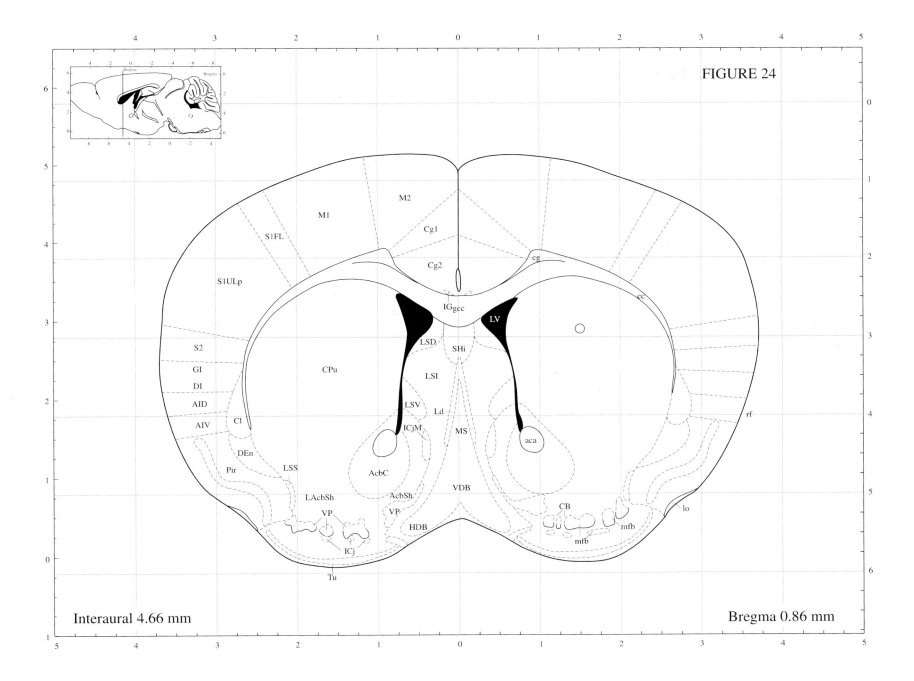

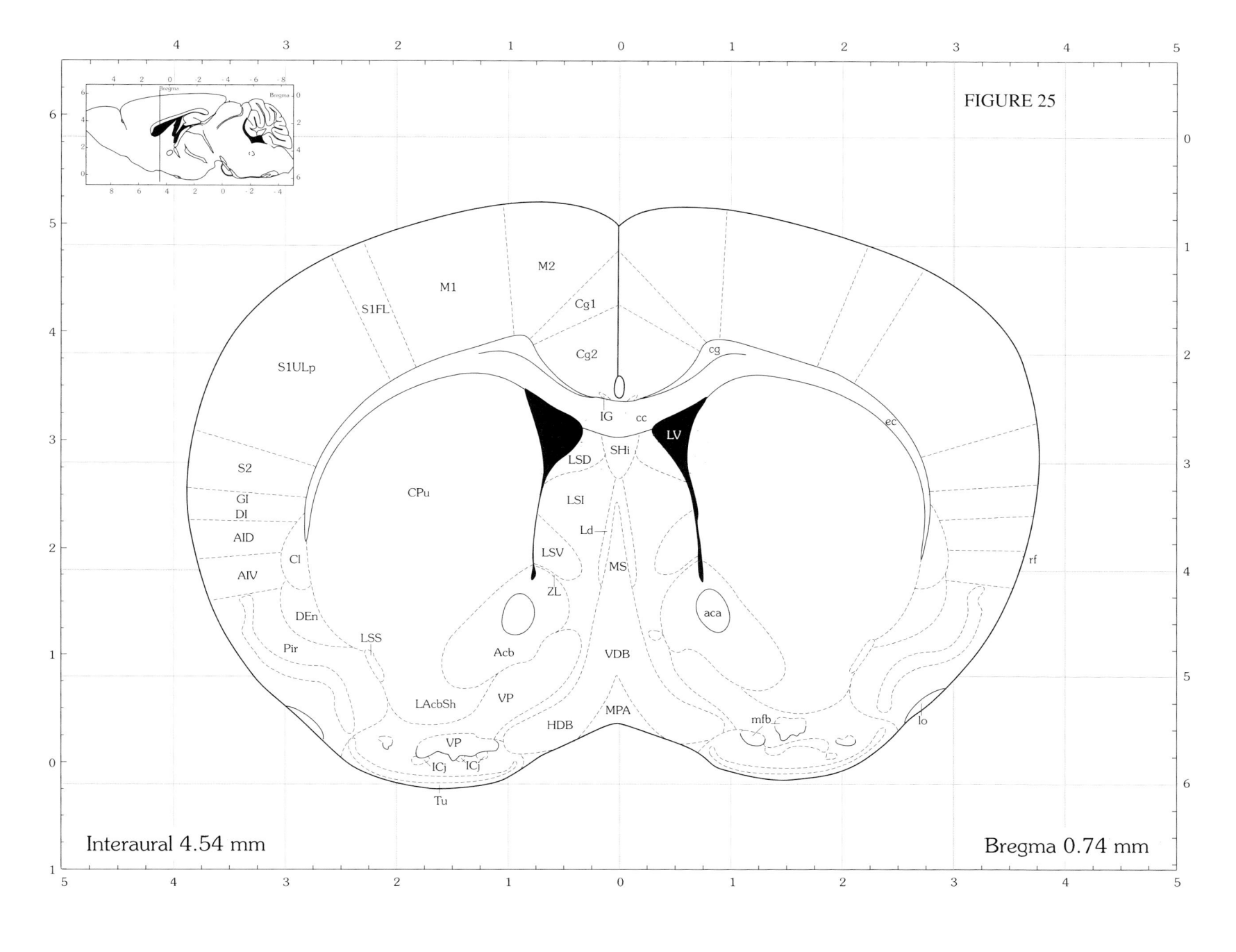

FIGURE 25

Interaural 4.54 mm

Bregma 0.74 mm

FIGURES 25 and 26
aca   ant comm, ant
Acb   accumbens nu
AID   agran insular cx dors

AIV   agran insular cx, vent
AVPe   antvent periventric nu
BST   bed nu, stria terminalis
BSTMA   BST, med div, ant

cc   corpus callosum
cg   cingulum
Cg1   cingulate cortex, area 1
Cg2   cingulate cortex, area 2

Cl   claustrum
CPu   caudate putamen
DEn   dors endopiriform nu
DI   dysgran insular cx

ec   ext capsule
GI   granular insular cx
HDB   nu horiz limb diag bd
ICj   isl, Calleja

IG   indusium griseum
Ld   lambdoid septal zone
lo   lat olfactory tr
LPO   lat preoptic area

LSD   lat septal nu, dors pt
LSI   lat septal nu, intermed
LSS   lat stripe, striatum
LSV   lat septal nu, vent pt

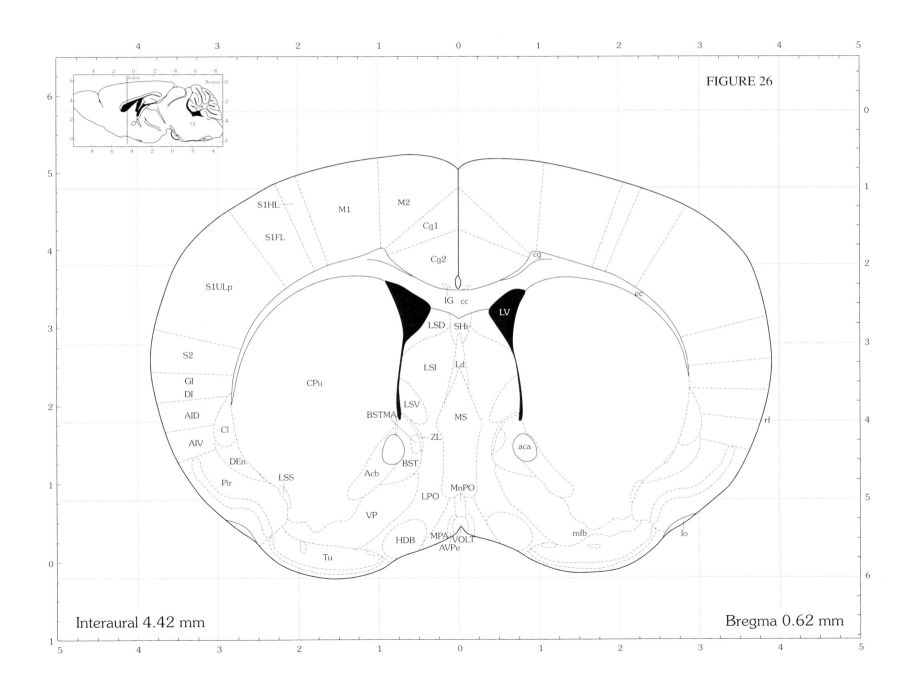

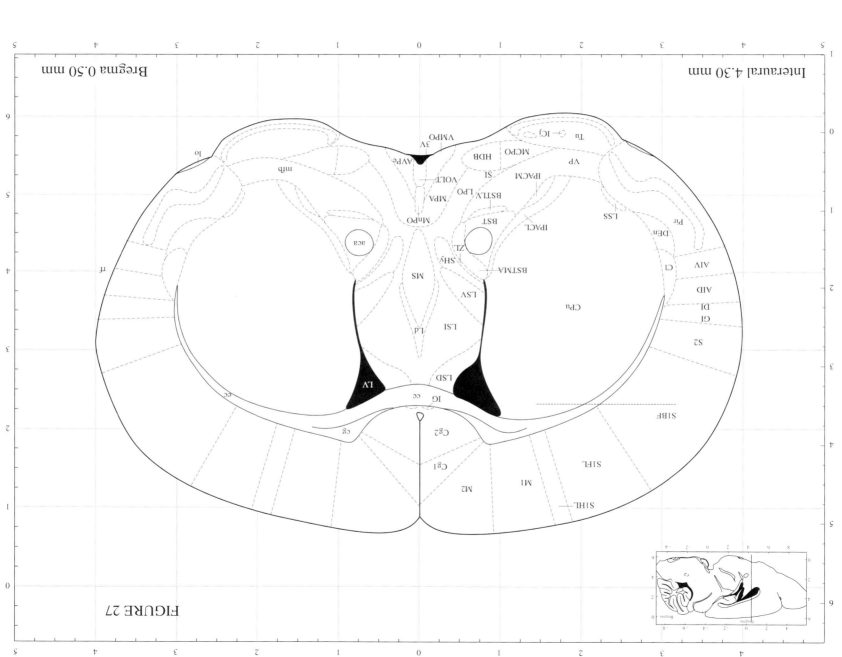

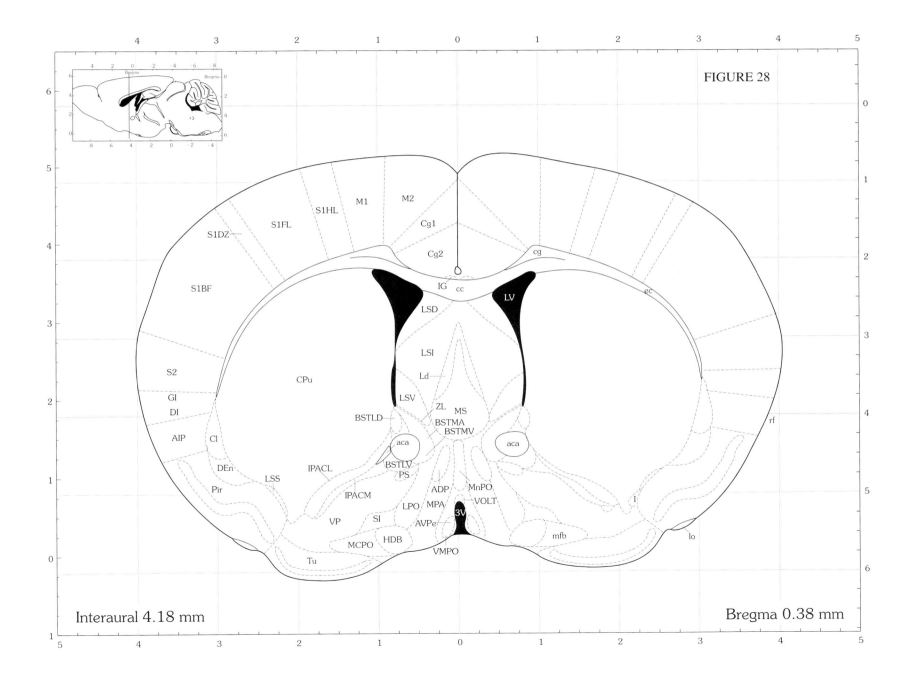

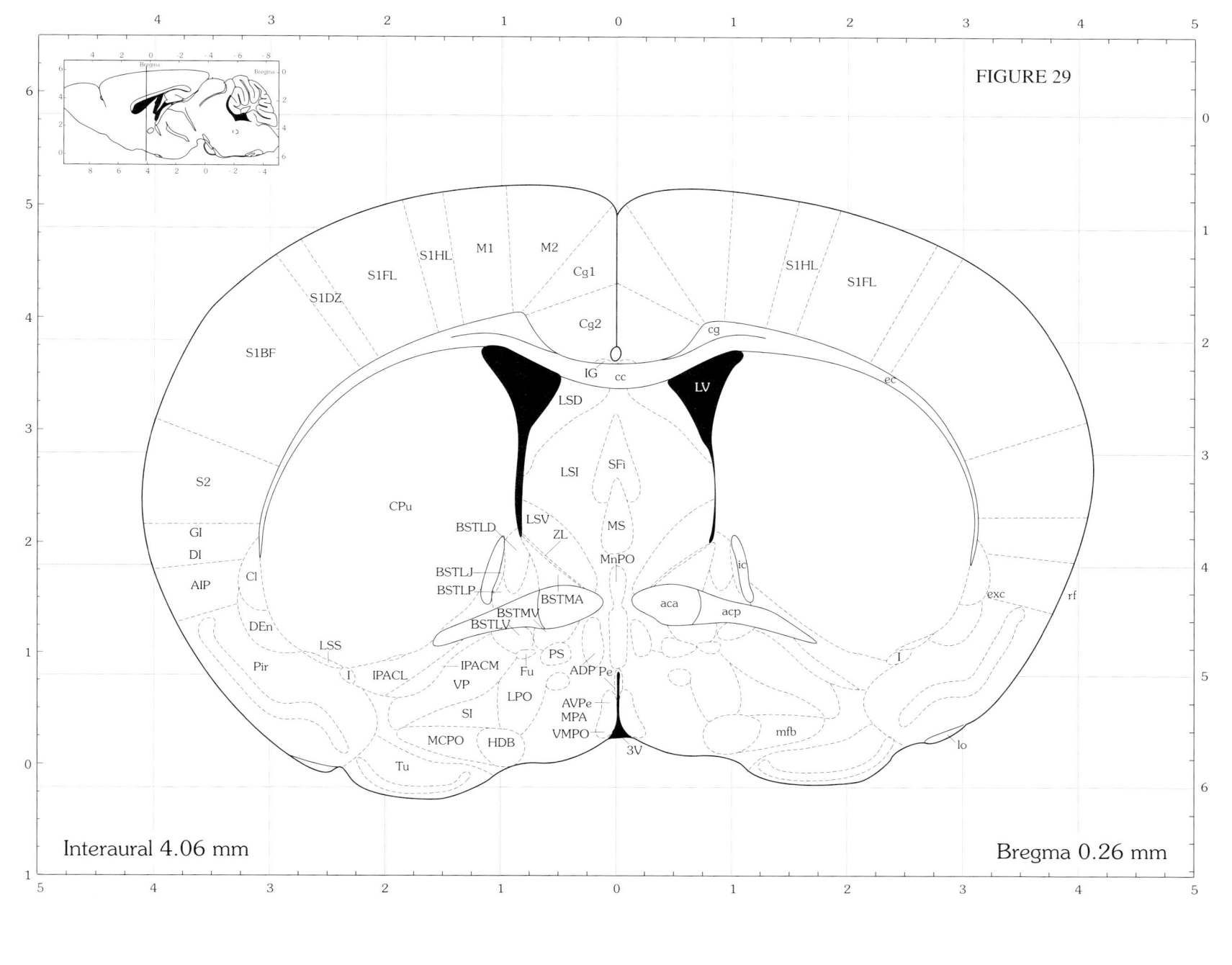

FIGURE 29

Interaural 4.06 mm

Bregma 0.26 mm

FIGURES 29 and 30
2n   optic nerve
3V   3rd ventric
aca   ant comm, ant

acp   ant comm, post
ADP   anterodors preoptic nu
AIP   agran insular cx, post
AVPe   antvent periventric nu

BSTLD   BST, lat div, dors pt
BSTLJ   BST lat div, juxtacap
BSTLP   BST, lat div, post
BSTLV   BST, lat div, vent

BSTMA   BST, med div, ant
BSTMV   BST, med div, vent
cc   corpus callosum
cg   cingulum

Cg1   cingulate cortex, area 1
Cg2   cingulate cortex, area 2
Cl   claustrum
CPu   caudate putamen

DEn   dors endopiriform nu
DI   dysgran insular cx
ec   ext capsule
exc   extreme capsule

f   fornix
Fu   BST, fusiform pt
GI   granular insular cx
HDB   nu horiz limb diag bd

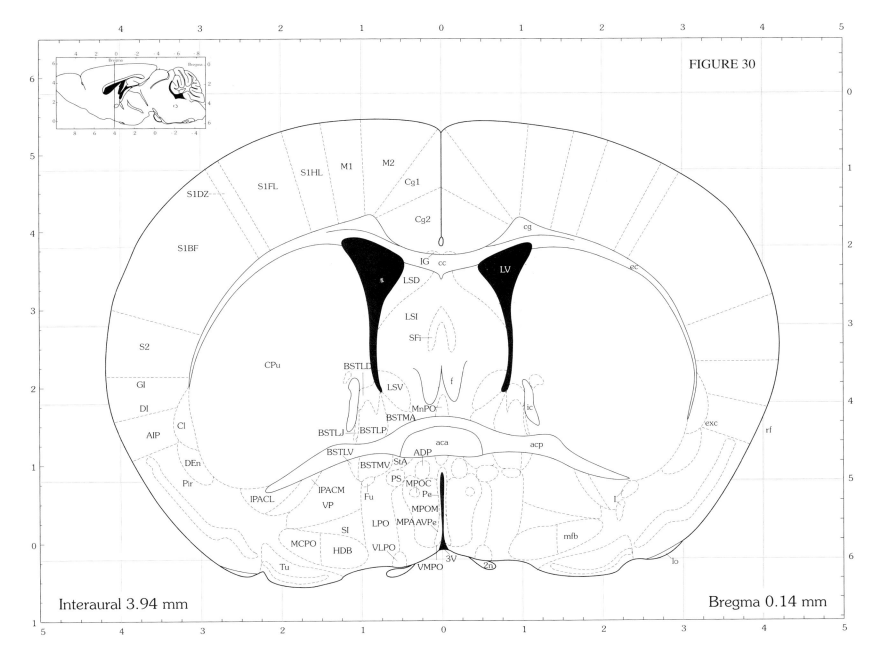

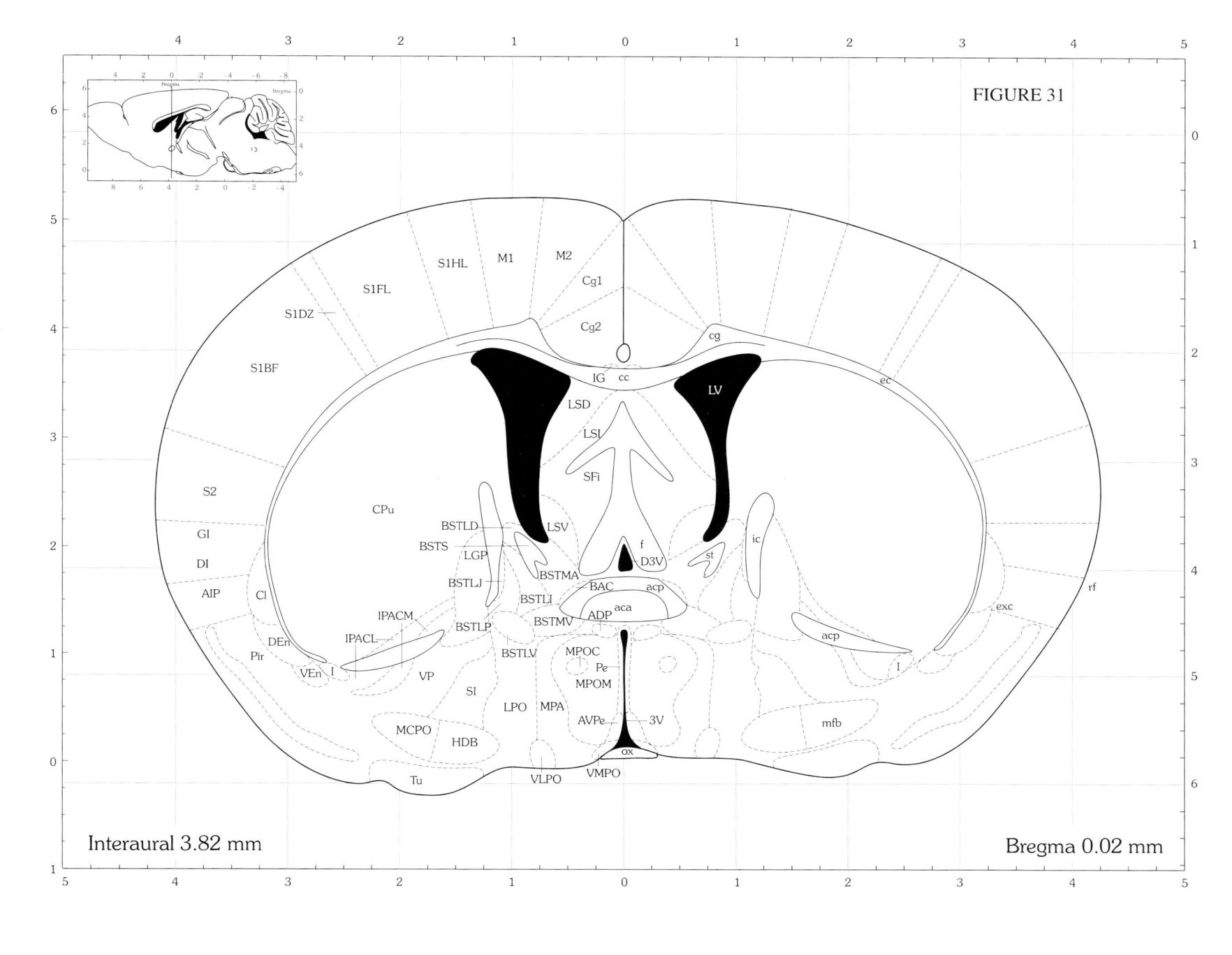

FIGURE 31

Interaural 3.82 mm

Bregma 0.02 mm

FIGURES 31 and 32
3V 3rd ventric
A14 A14 dopamine cells
AAD ant amyg area, dors

AAV ant amyg area, vent
aca ant comm, ant
acp ant comm, post
ADP anterodors preoptic nu

AIP agran insular cx post
BAC bed nu, ant comm
BSTLI BST, lat div, intermed
BSTLP BST, lat div, post

BSTMPI BST med, post-interm
BSTMPM BST med div post m
BSTMV BST, med div, vent
BSTS BST, supracap pt

cc corpus callosum
cg cingulum
Cg1 cingulate cortex, area 1
Cg2 cingulate cortex, area 2

Cl claustrum
CPu caudate putamen
CxA cortex-amyg insular cx zn
D3V dors 3rd ventric

DEn dors endopiriform nu
DI dysgran insular cx
ec ext capsule
exc extreme capsule

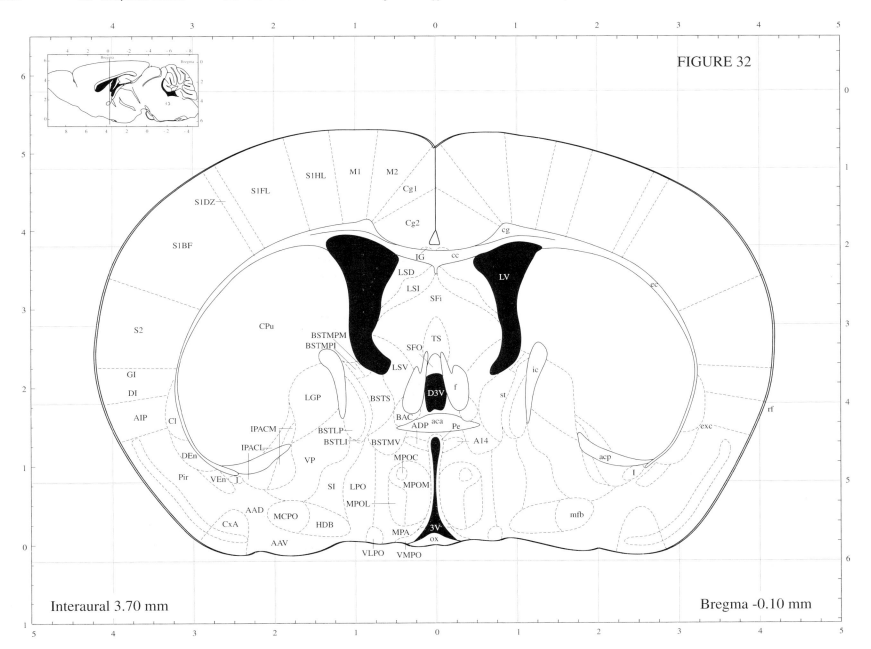

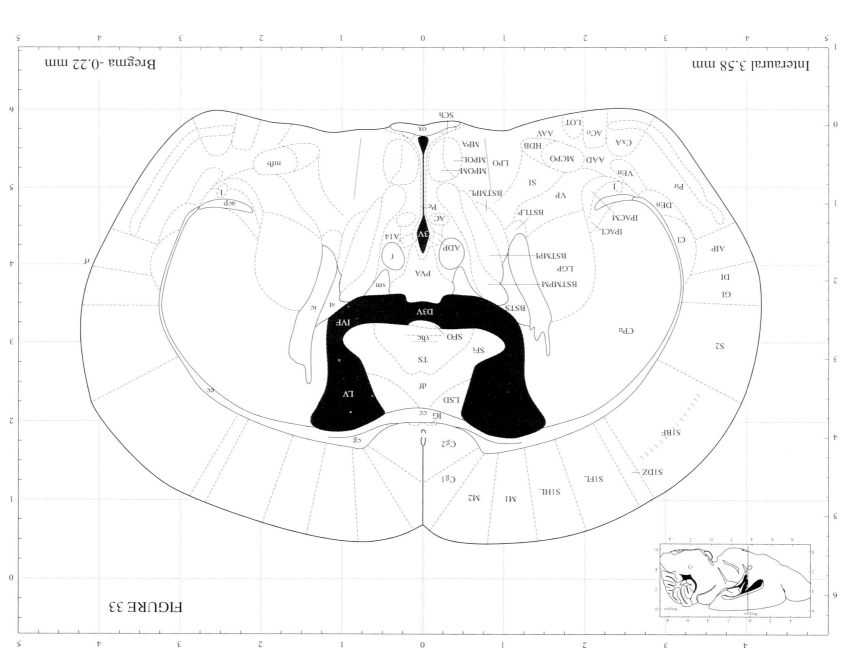

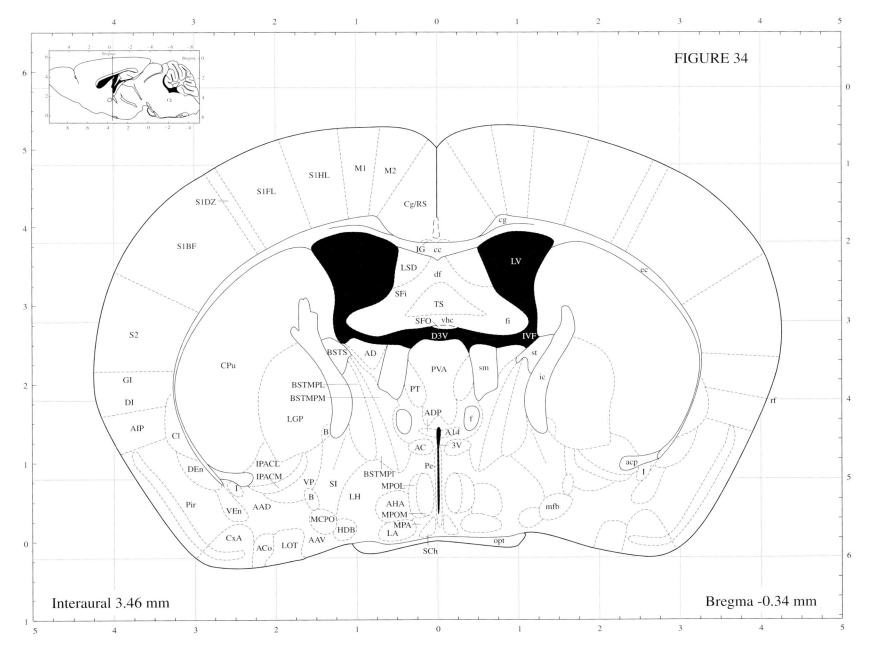

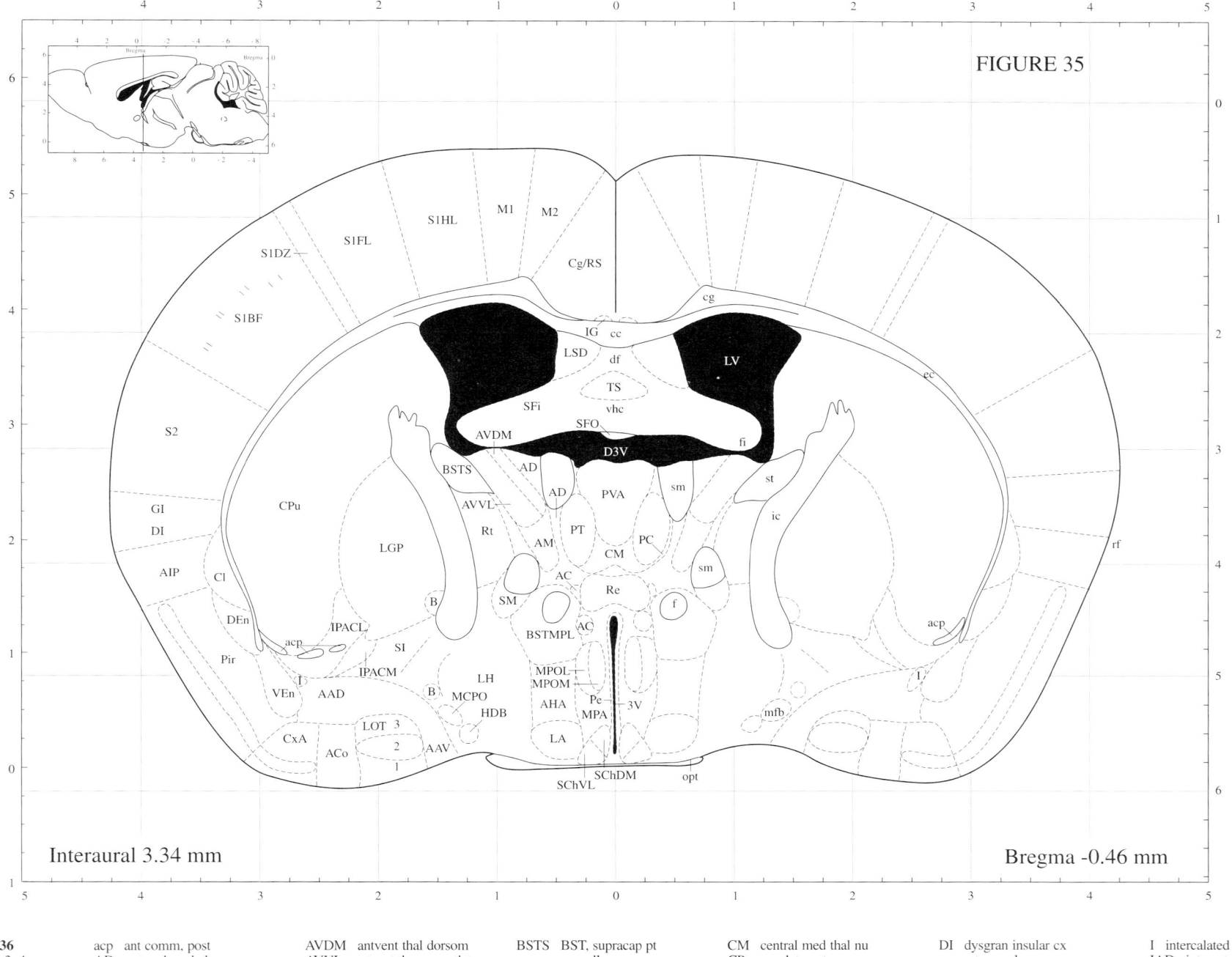

FIGURE 35

Interaural 3.34 mm

Bregma -0.46 mm

FIGURES 35 and 36

l, 2, 3, 4 layer 1, 2, 3, 4
3V 3rd ventric
AAD ant amyg area, dors
AAV ant amyg area, vent
ACo ant cortical amyg nu

acp ant comm, post
AD anterodors thal nu
AHA ant hypothal area, ant
AIP agran insular cx post
AM anteromed thal nu
AMV anteromed th nu, vent

AVDM antvent thal dorsom
AVVL antvent th nu, ventlat
B basal nu (Meynert)
BLA basolat amyg nu, ant
BMA basmed amyg nu, ant
BSTMPL BST, med, posterolat

BSTS BST, supracap pt
cc corpus callosum
CeM ce amyg nu, med div
cg cingulum
Cg/RS cingulate/retrosplenial
Cl claustrum

CM central med thal nu
CPu caudate putamen
CxA cortex-amyg transit zn
D3V dors 3rd ventric
DEn dors endopiriform nu
df dors fornix

DI dysgran insular cx
ec ext capsule
f fornix
fi fimbria, hippocampus
GI granular insular cx
HDB nu horiz limb diag bd

I intercalated nuclei, amyg
IAD interanterodors thal nu
ic int capsule
IG indusium griseum
IPACL IPAC, lat
IPACM IPAC, med

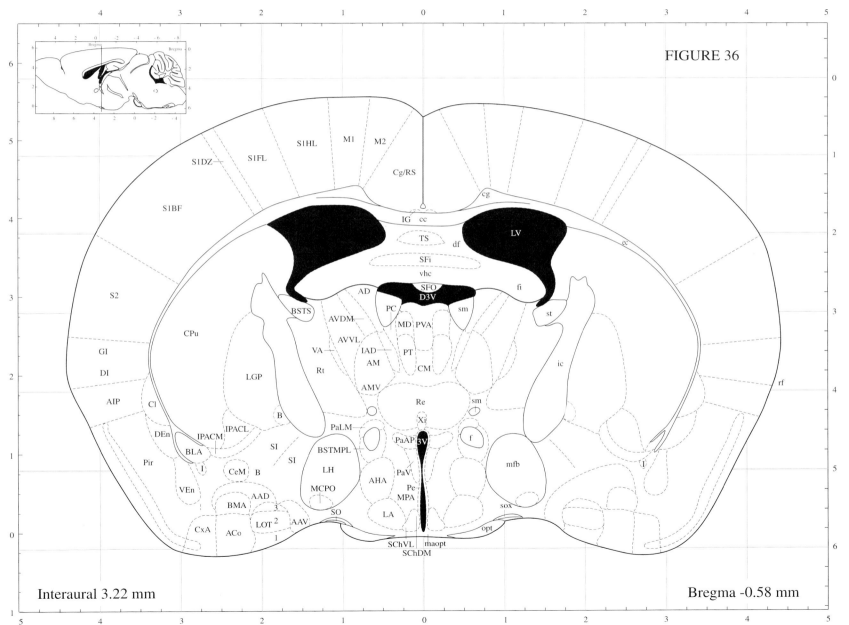

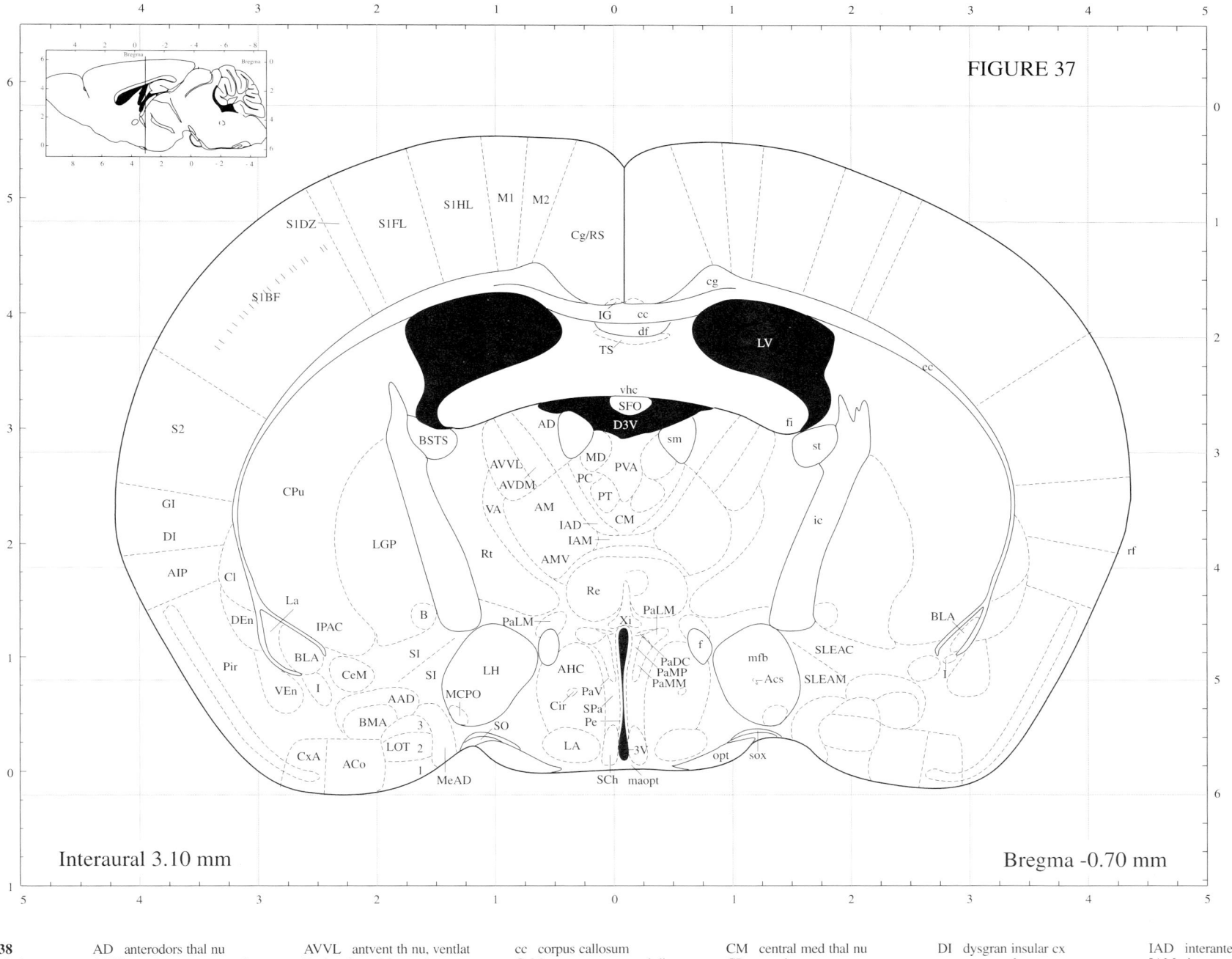

FIGURE 37

Interaural 3.10 mm

Bregma -0.70 mm

FIGURES 37 and 38
l, 2, 3, 4 layer 1, 2, 3, 4
3V 3rd ventric
AAD ant amyg area, dors
ACo ant cortical amyg nu
Acs accessory nu vent horn

AD anterodors thal nu
AHC ant hyp area, central
AIP agran insular cx post
AM anteromed thal nu
AMV anteromed th nu, vent
AVDM antvent thal dorsom

AVVL antvent th nu, ventlat
B basal nu (Meynert)
BAOT bed nu, acc olf tr
BLA basolat amyg nu, ant
BMA basmed amyg nu, ant
BSTS BST, supracap pt

cc corpus callosum
CeM ce amyg nu, med div
cg cingulum
Cg/RS cingulate/retrosplenial
Cir circular nu
Cl claustrum

CM central med thal nu
CPu caudate putamen
CxA cortex-amyg transit zn
D3V dors 3rd ventric
DEn dors endopiriform nu
df dors fornix

DI dysgran insular cx
ec ext capsule
f fornix
fi fimbria, hippocampus
GI granular insular cx
I intercalated nuclei, amyg

IAD interanterodors thal nu
IAM interanteromed thal nu
ic int capsule
IG indusium griseum
IPAC interstit nu, p limb, ac
LA lateroant hypothal nu

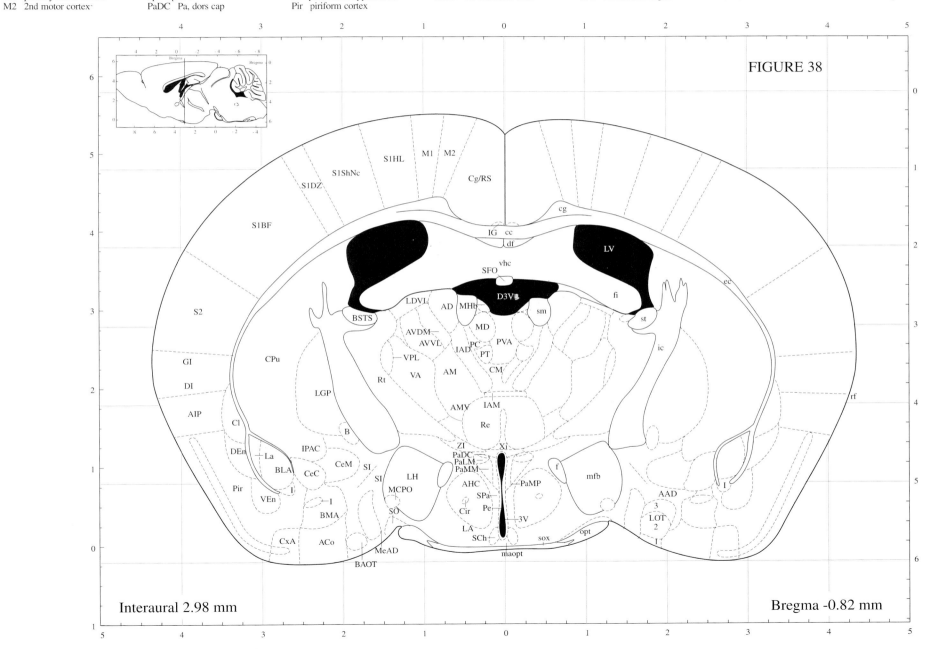

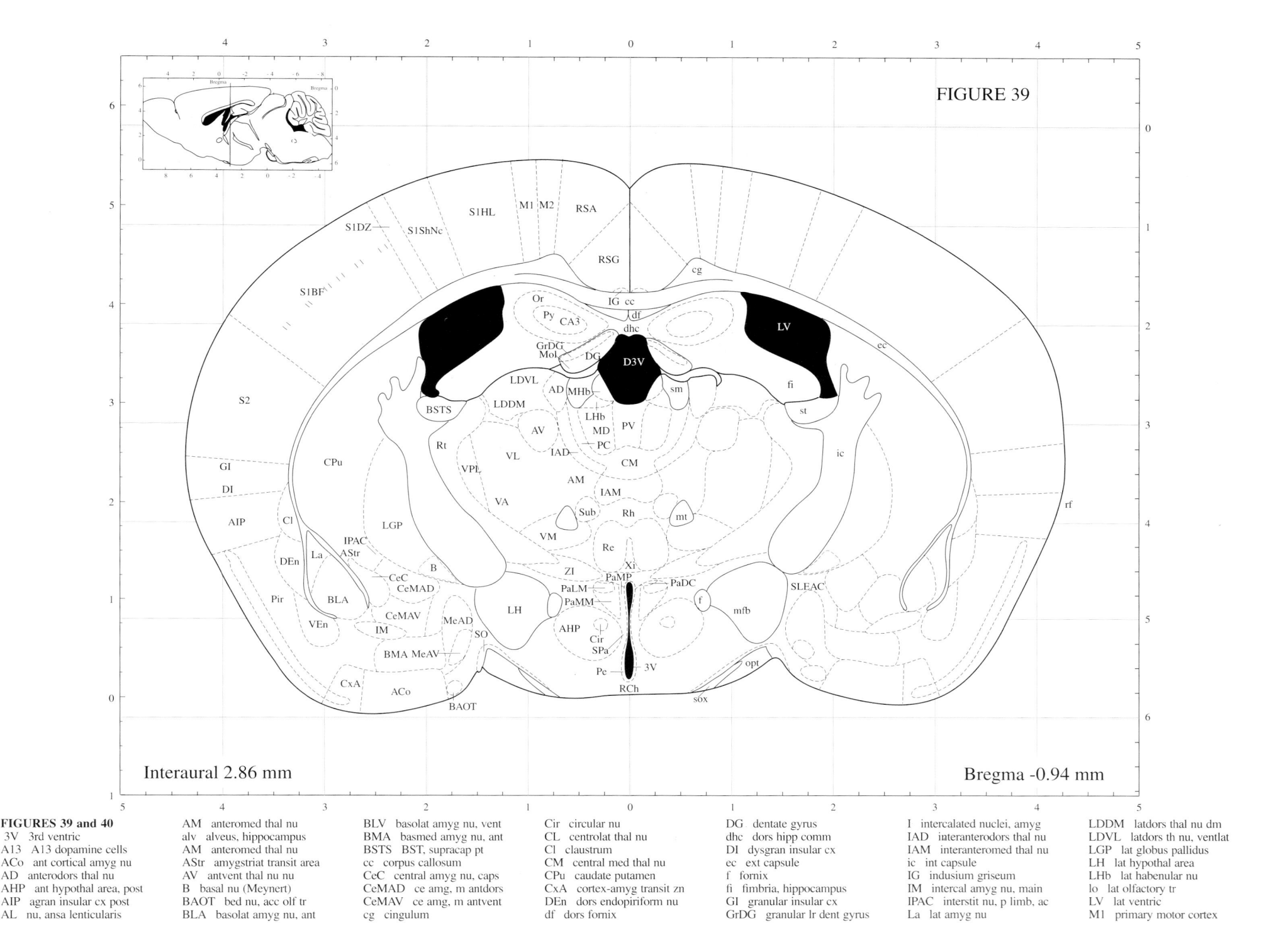

FIGURE 39

Interaural 2.86 mm

Bregma -0.94 mm

FIGURES 39 and 40

3V 3rd ventric
A13 A13 dopamine cells
ACo ant cortical amyg nu
AD anterodors thal nu
AHP ant hypothal area, post
AIP agran insular cx post
AL nu, ansa lenticularis

AM anteromed thal nu
alv alveus, hippocampus
AM anteromed thal nu
AStr amygstriat transit area
AV antvent thal nu nu
B basal nu (Meynert)
BAOT bed nu, acc olf tr
BLA basolat amyg nu, ant

BLV basolat amyg nu, vent
BMA basmed amyg nu, ant
BSTS BST, supracap pt
cc corpus callosum
CeC central amyg nu, caps
CeMAD ce amg, m antdors
CeMAV ce amg, m antvent
cg cingulum

Cir circular nu
CL centrolat thal nu
Cl claustrum
CM central med thal nu
CPu caudate putamen
CxA cortex-amyg transit zn
DEn dors endopiriform nu
df dors fornix

DG dentate gyrus
dhc dors hipp comm
DI dysgran insular cx
ec ext capsule
f fornix
fi fimbria, hippocampus
GI granular insular cx
GrDG granular lr dent gyrus

I intercalated nuclei, amyg
IAD interanterodors thal nu
IAM interanteromed thal nu
ic int capsule
IG indusium griseum
IM intercal amyg nu, main
IPAC interstit nu, p limb, ac
La lat amyg nu

LDDM latdors thal nu dm
LDVL latdors th nu, ventlat
LGP lat globus pallidus
LH lat hypothal area
LHb lat habenular nu
lo lat olfactory tr
LV lat ventric
M1 primary motor cortex

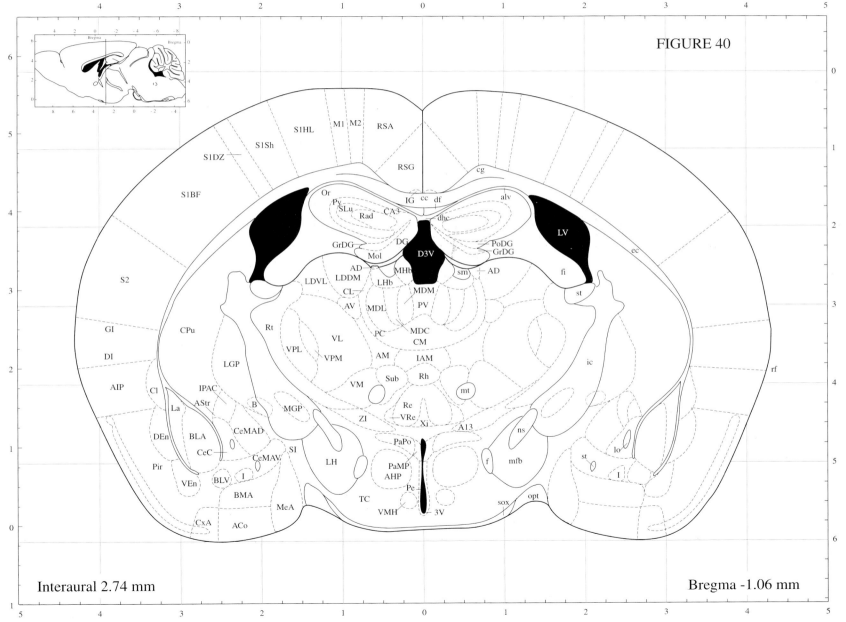

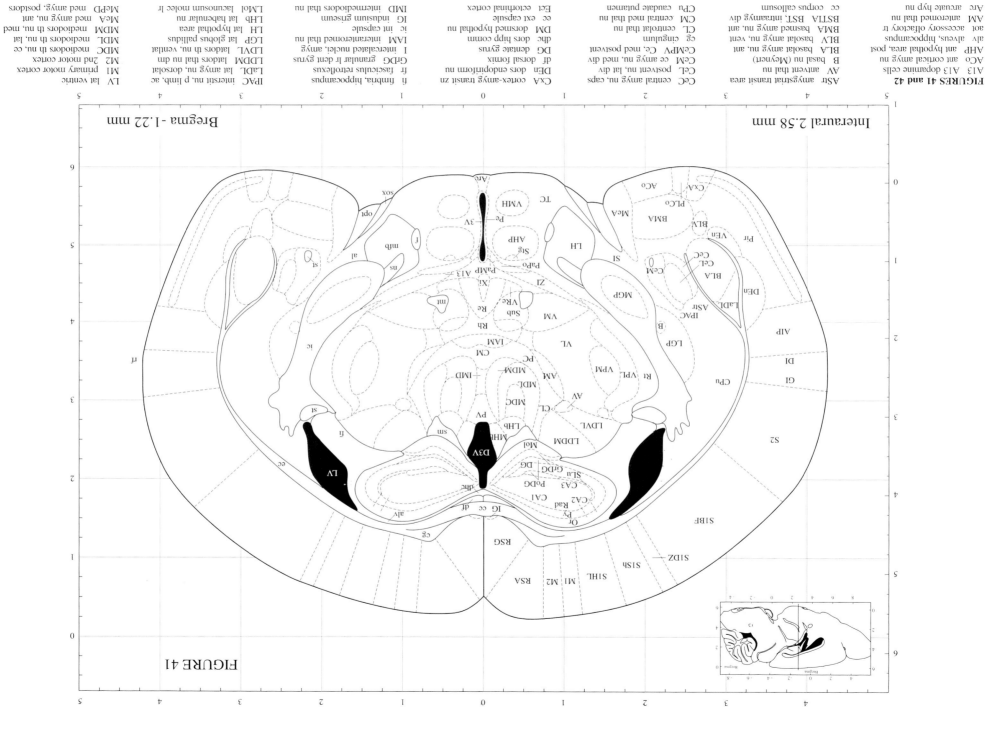

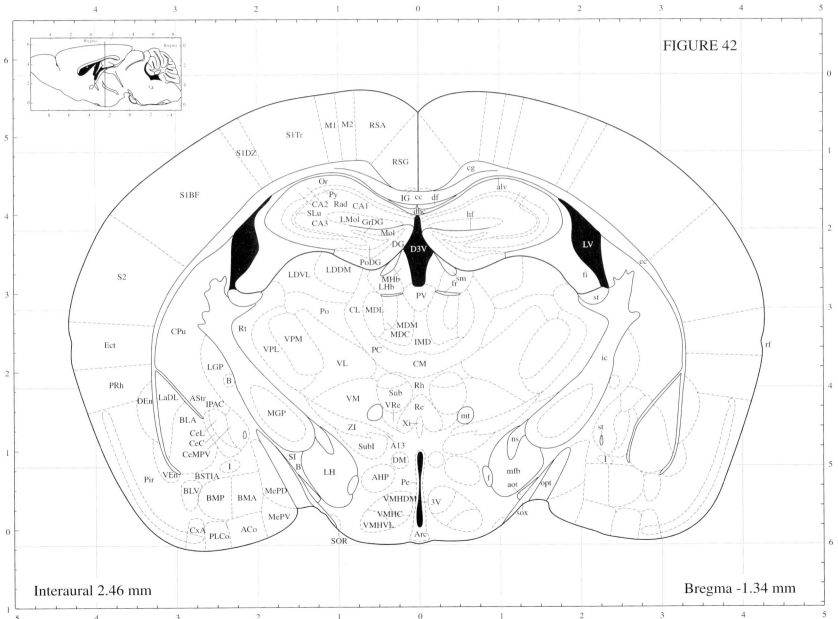

FIGURE 42

Interaural 2.46 mm

Bregma -1.34 mm

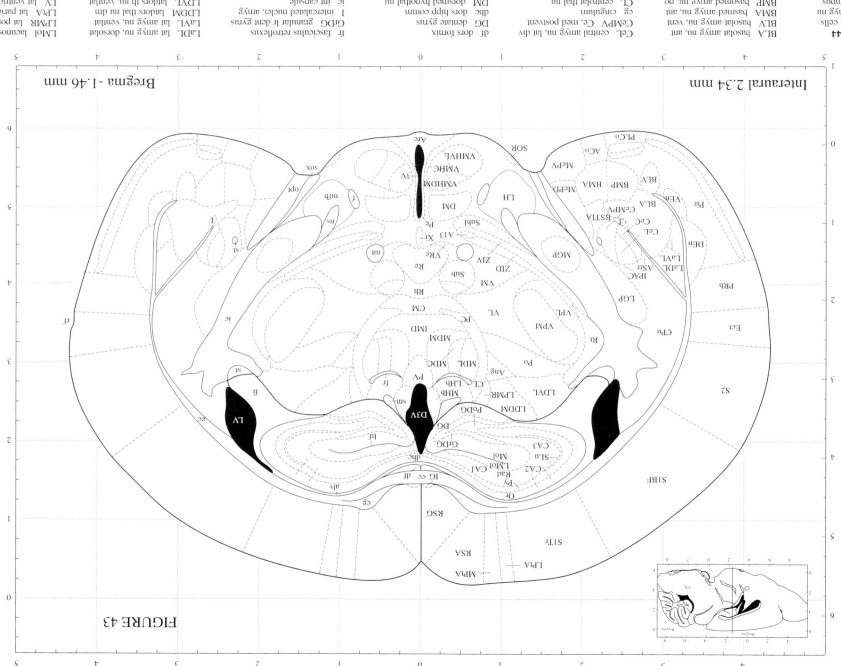

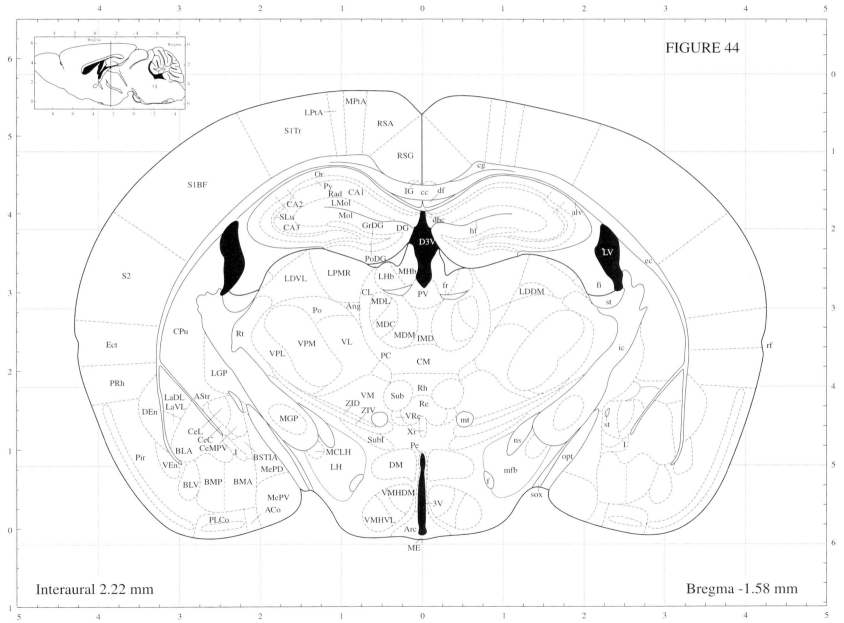

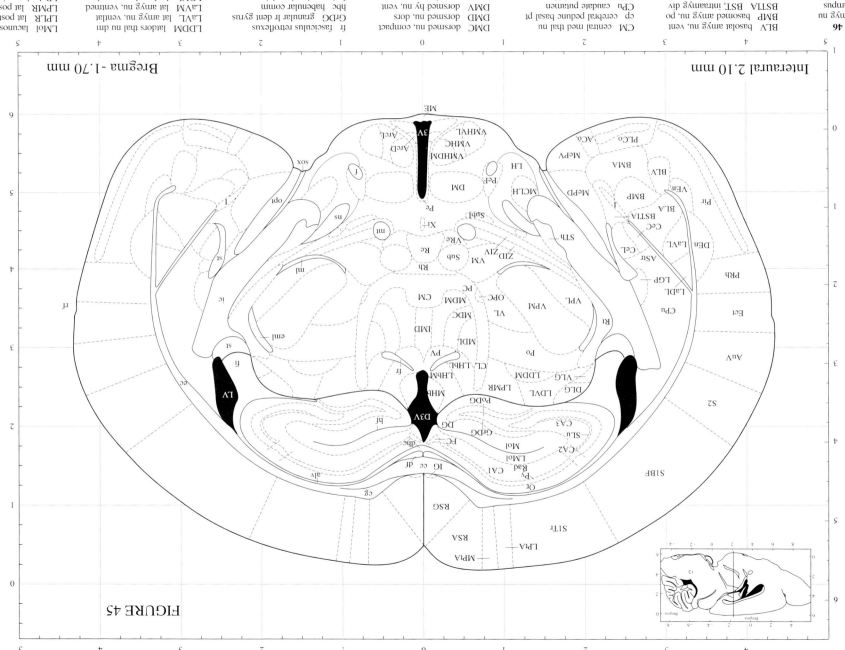

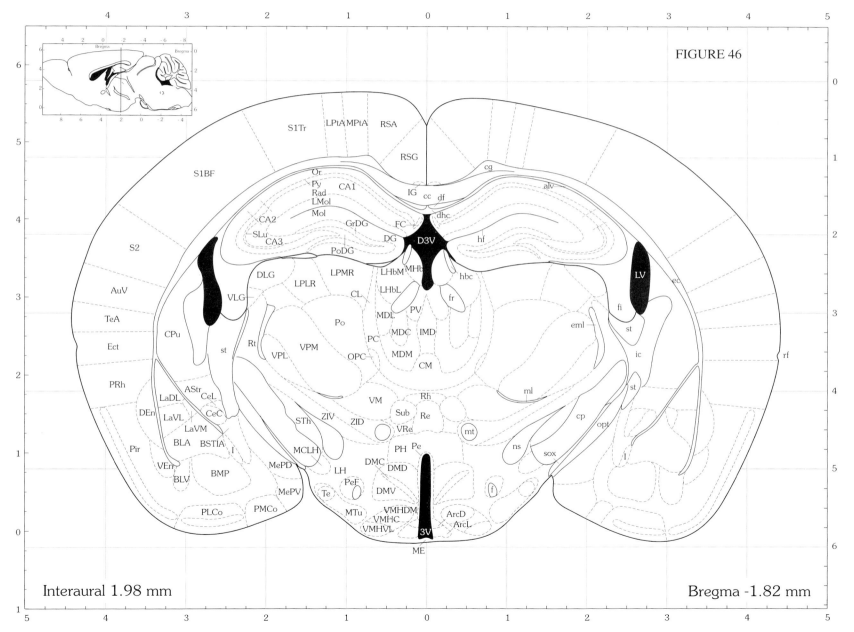

FIGURE 46

Interaural 1.98 mm　　Bregma −1.82 mm

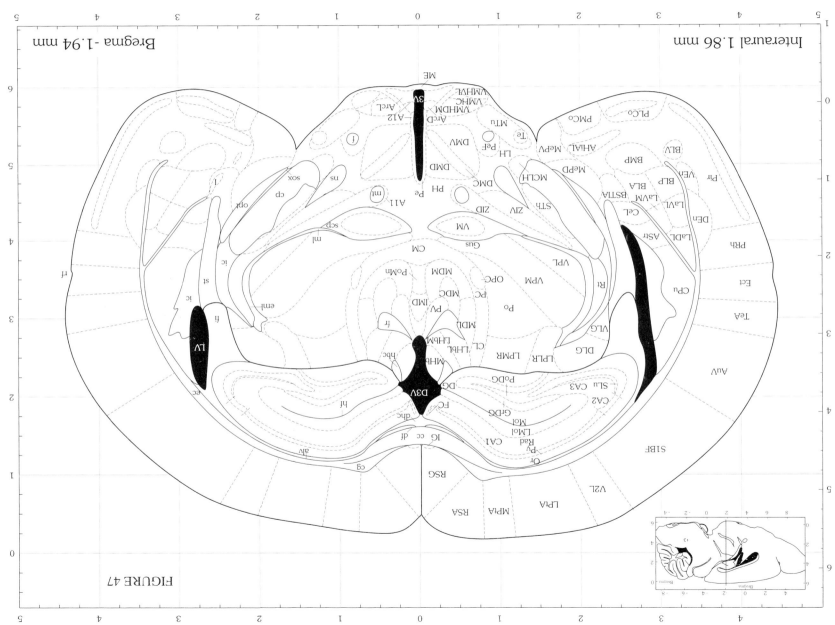

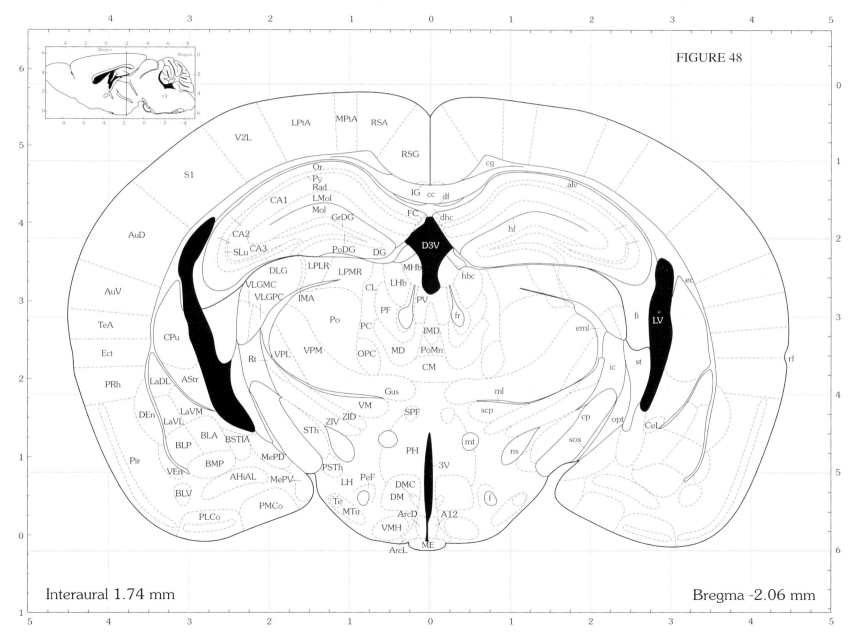

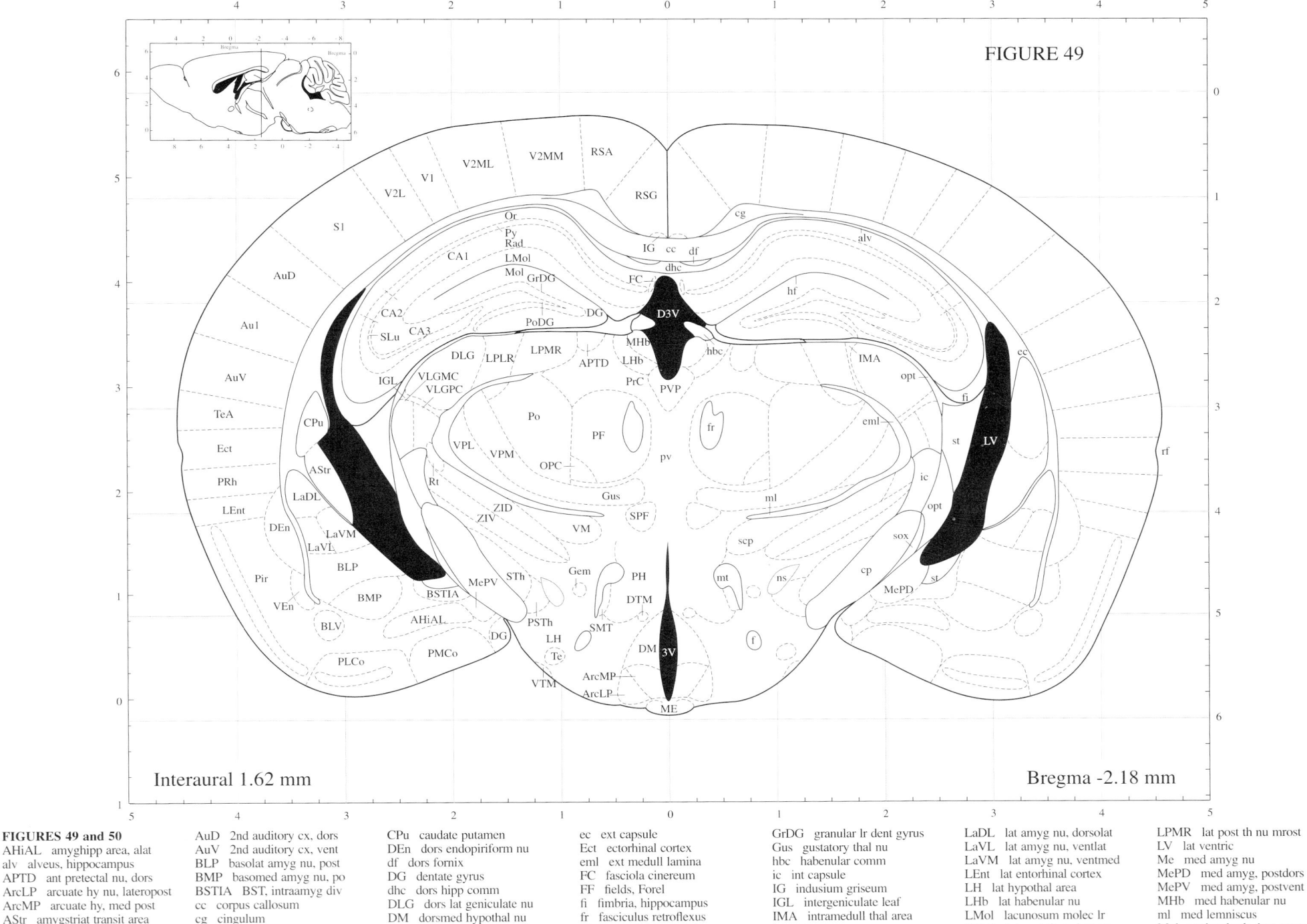

FIGURE 49

Interaural 1.62 mm

Bregma -2.18 mm

FIGURES 49 and 50

AHiAL amyghipp area, alat
alv alveus, hippocampus
APTD ant pretectal nu, dors
ArcLP arcuate hy nu, lateropost
ArcMP arcuate hy, med post
AStr amygstriat transit area
Au1 primary auditory cortex

AuD 2nd auditory cx, dors
AuV 2nd auditory cx, vent
BLP basolat amyg nu, post
BMP basomed amyg nu, po
BSTIA BST, intraamyg div
cc corpus callosum
cg cingulum
cp cereb pedunc basal pt

CPu caudate putamen
DEn dors endopiriform nu
df dors fornix
DG dentate gyrus
dhc dors hipp comm
DLG dors lat geniculate nu
DM dorsmed hypothal nu
DTM dors tuberomammill

ec ext capsule
Ect ectorhinal cortex
eml ext medull lamina
FC fasciola cinereum
FF fields, Forel
fi fimbria, hippocampus
fr fasciculus retroflexus
Gem gemini hypothal nu

GrDG granular lr dent gyrus
Gus gustatory thal nu
hbc habenular comm
ic int capsule
IG indusium griseum
IGL intergeniculate leaf
IMA intramedull thal area
La lat amyg nu

LaDL lat amyg nu, dorsolat
LaVL lat amyg nu, ventlat
LaVM lat amyg nu, ventmed
LEnt lat entorhinal cortex
LH lat hypothal area
LHb lat habenular nu
LMol lacunosum molec lr
LPLR lat post th nu latpost

LPMR lat post th nu mrost
LV lat ventric
Me med amyg nu
MePD med amyg, postdors
MePV med amyg, postvent
MHb med habenular nu
ml med lemniscus
Mol molecular lr dent gyrus

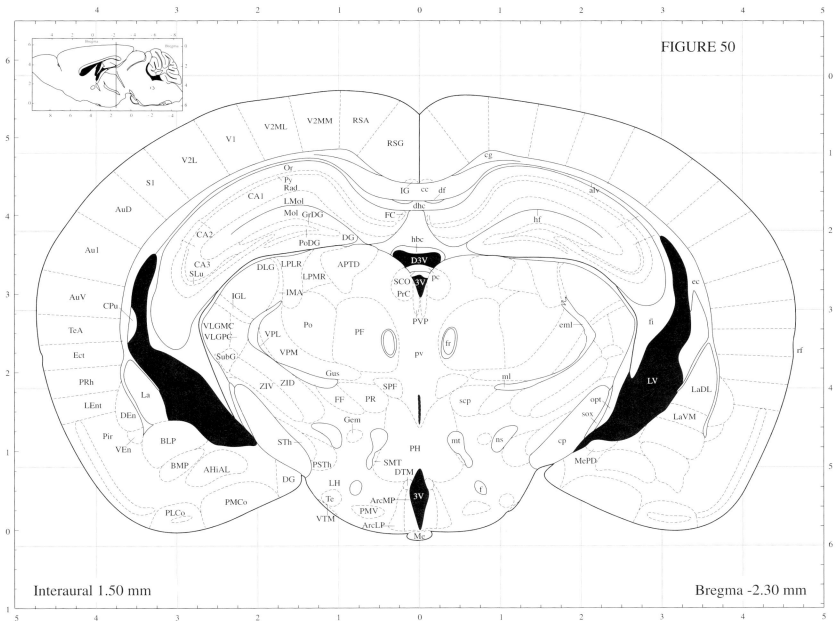

FIGURE 50

Interaural 1.50 mm

Bregma -2.30 mm

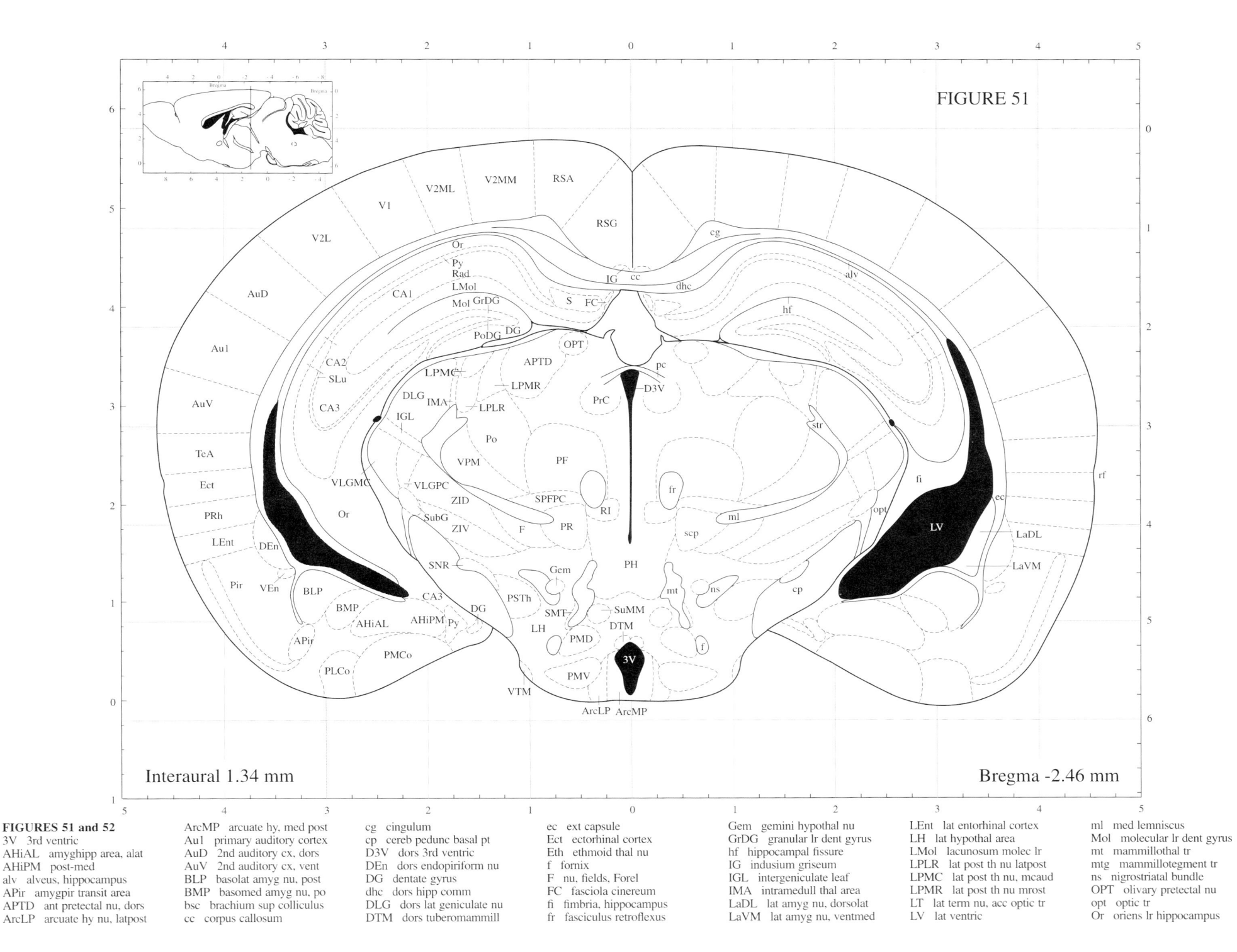

FIGURE 51

Interaural 1.34 mm

Bregma -2.46 mm

FIGURES 51 and 52
3V   3rd ventric
AHiAL   amyghipp area, alat
AHiPM   post-med
alv   alveus, hippocampus
APir   amygpir transit area
APTD   ant pretectal nu, dors
ArcLP   arcuate hy nu, latpost

ArcMP   arcuate hy, med post
Au1   primary auditory cortex
AuD   2nd auditory cx, dors
AuV   2nd auditory cx, vent
BLP   basolat amyg nu, post
BMP   basomed amyg nu, po
bsc   brachium sup colliculus
cc   corpus callosum

cg   cingulum
cp   cereb pedunc basal pt
D3V   dors 3rd ventric
DEn   dors endopiriform nu
DG   dentate gyrus
dhc   dors hipp comm
DLG   dors lat geniculate nu
DTM   dors tuberomammill

ec   ext capsule
Ect   ectorhinal cortex
Eth   ethmoid thal nu
f   fornix
F   nu, fields, Forel
FC   fasciola cinereum
fi   fimbria, hippocampus
fr   fasciculus retroflexus

Gem   gemini hypothal nu
GrDG   granular lr dent gyrus
hf   hippocampal fissure
IG   indusium griseum
IGL   intergeniculate leaf
IMA   intramedull thal area
LaDL   lat amyg nu, dorsolat
LaVM   lat amyg nu, ventmed

LEnt   lat entorhinal cortex
LH   lat hypothal area
LMol   lacunosum molec lr
LPLR   lat post th nu latpost
LPMC   lat post th nu, mcaud
LPMR   lat post th nu mrost
LT   lat term nu, acc optic tr
LV   lat ventric

ml   med lemniscus
Mol   molecular lr dent gyrus
mt   mammillothal tr
mtg   mammillotegment tr
ns   nigrostriatal bundle
OPT   olivary pretectal nu
opt   optic tr
Or   oriens lr hippocampus

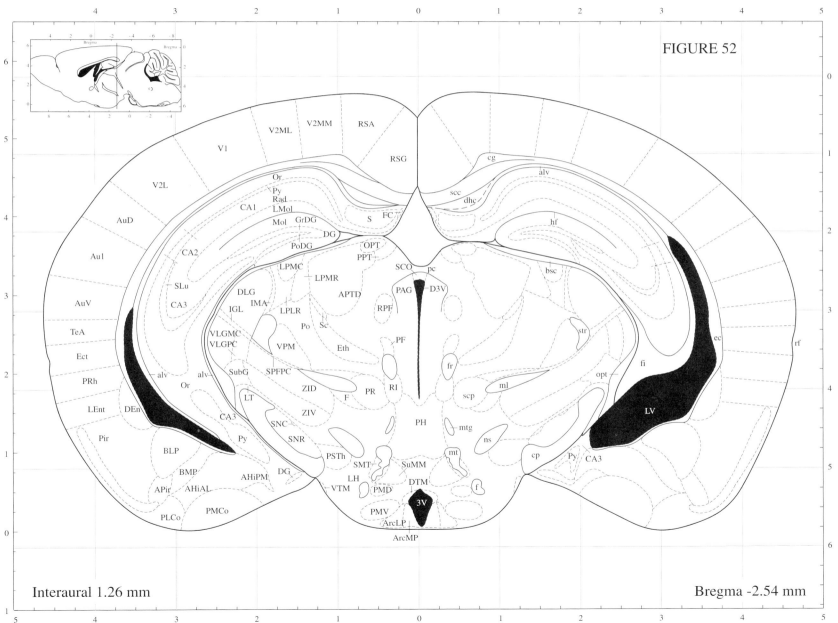

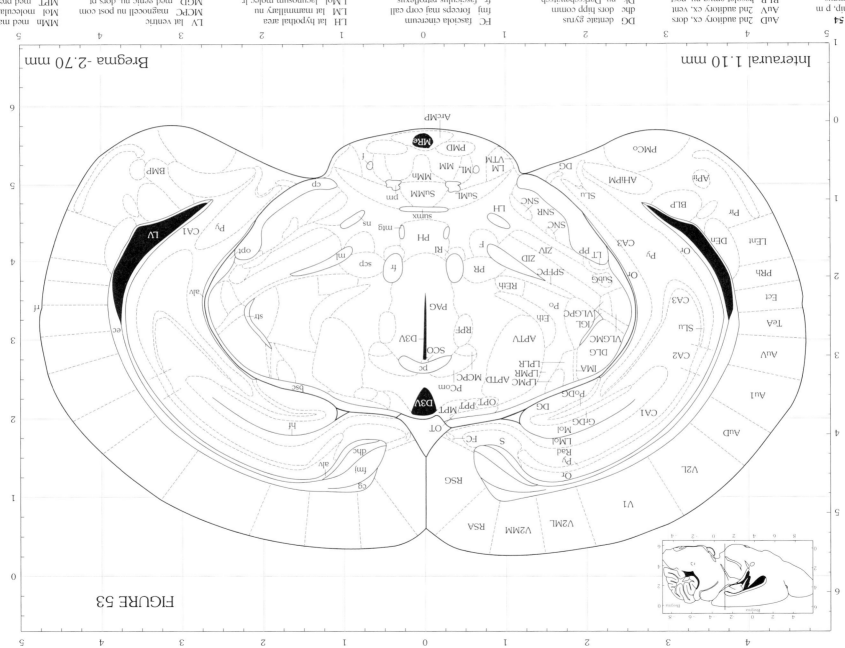

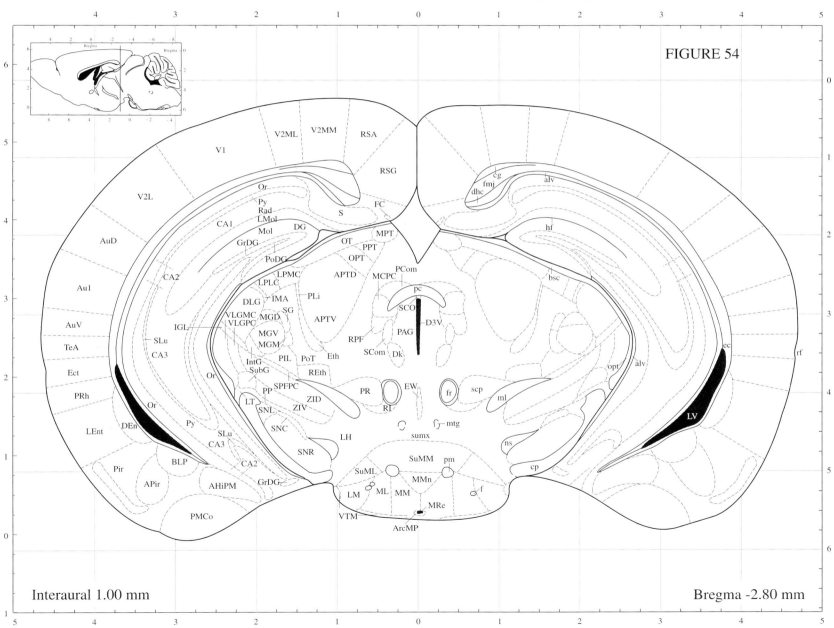

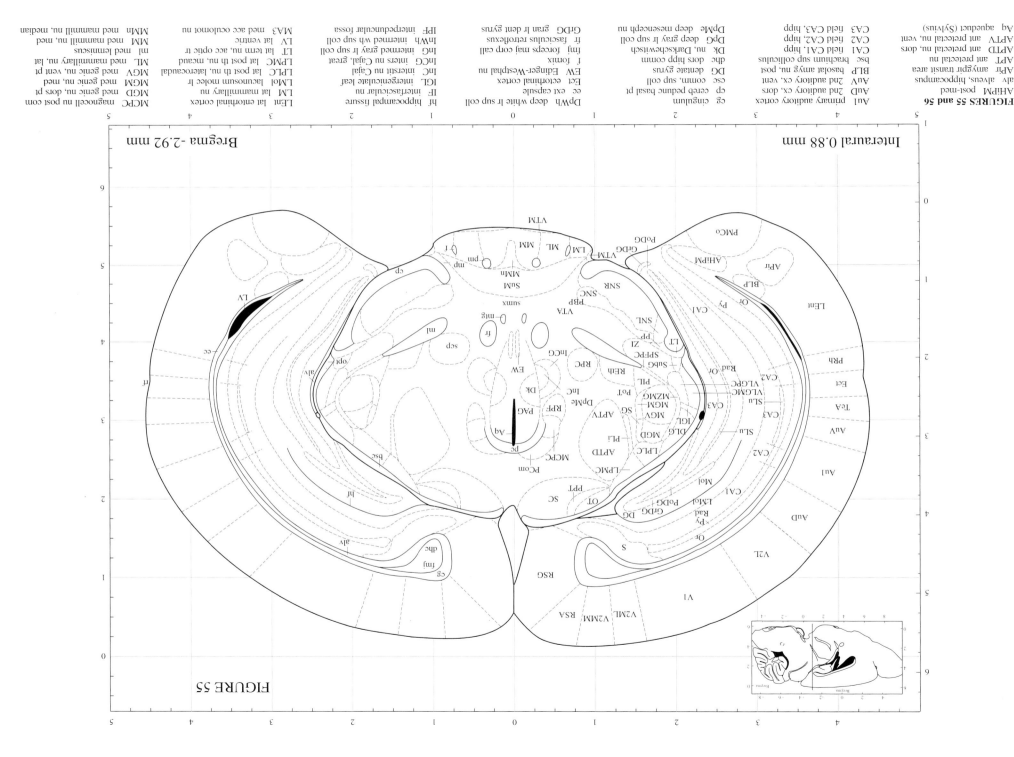

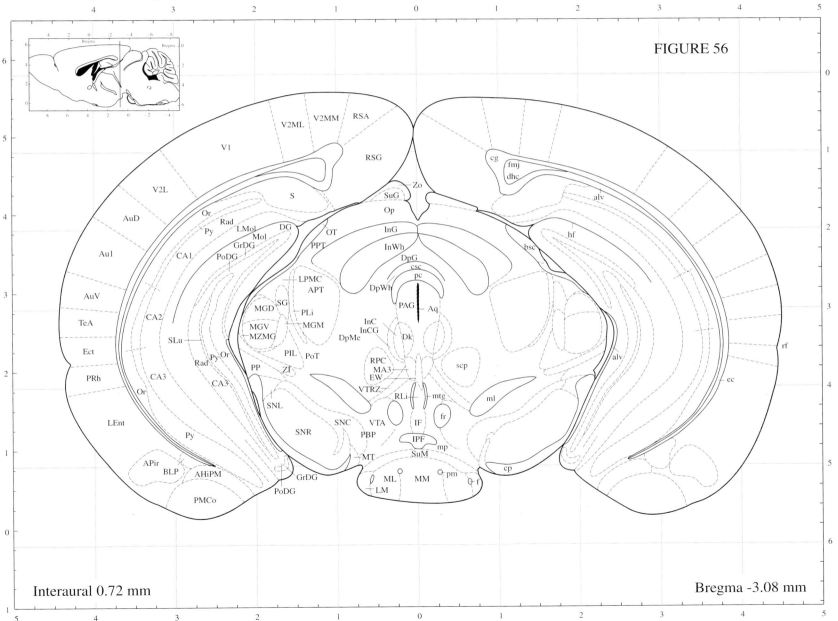

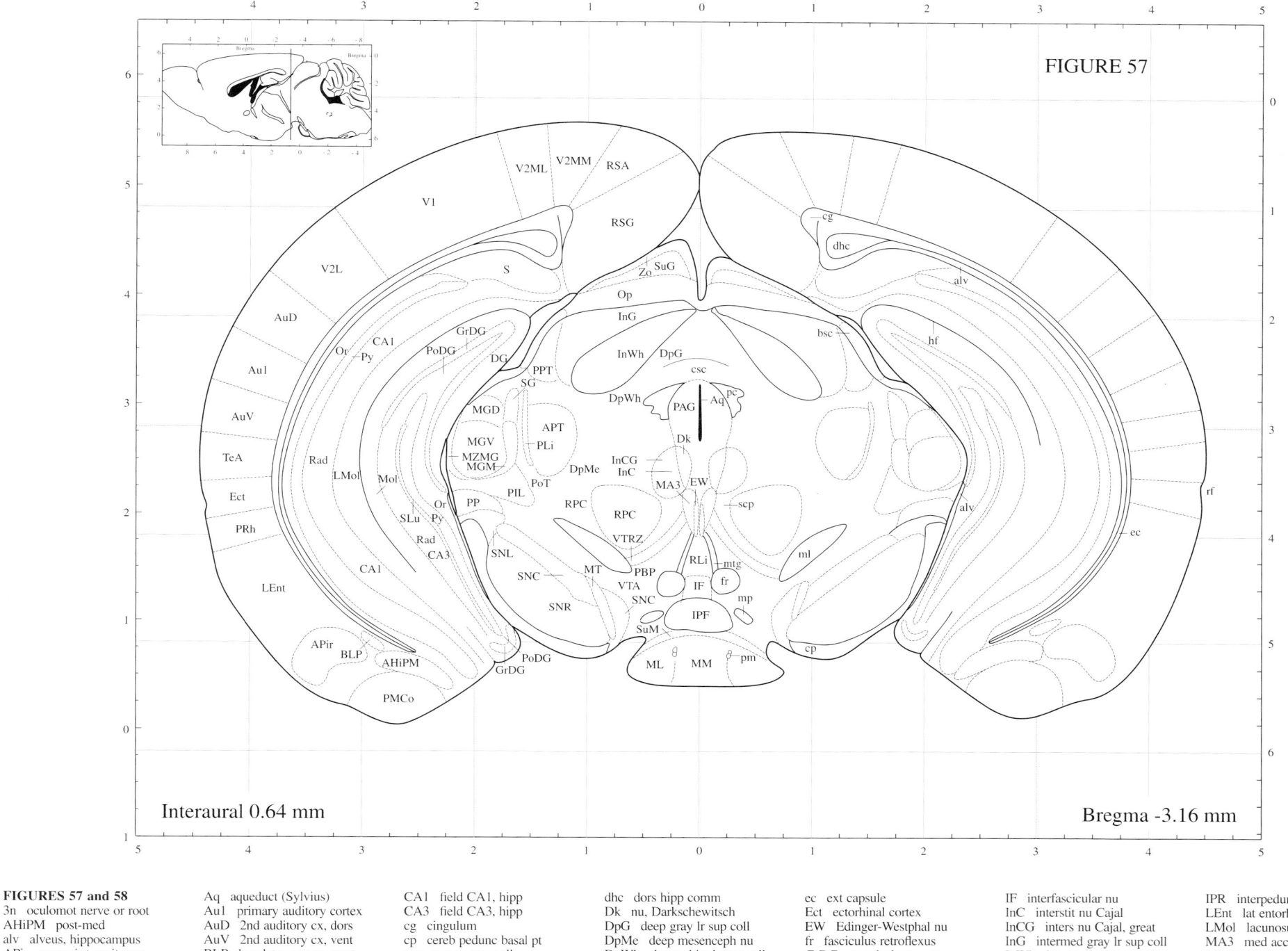

FIGURE 57

Interaural 0.64 mm

Bregma -3.16 mm

FIGURES 57 and 58

3n oculomot nerve or root
AHiPM post-med
alv alveus, hippocampus
APir amygpir transit area
APT ant pretectal nu

Aq aqueduct (Sylvius)
Au1 primary auditory cortex
AuD 2nd auditory cx, dors
AuV 2nd auditory cx, vent
BLP basolat amyg nu, post
bsc brachium sup colliculus

CA1 field CA1, hipp
CA3 field CA3, hipp
cg cingulum
cp cereb pedunc basal pt
csc comm, sup coll
DG dentate gyrus

dhc dors hipp comm
Dk nu, Darkschewitsch
DpG deep gray lr sup coll
DpMe deep mesenceph nu
DpWh deep white lr sup coll
DT dors term nu acc opt tr

ec ext capsule
Ect ectorhinal cortex
EW Edinger-Westphal nu
fr fasciculus retroflexus
GrDG gran lr dent gyrus
hf hippocampal fissure

IF interfascicular nu
InC interstit nu Cajal
InCG inters nu Cajal, great
InG intermed gray lr sup coll
InWh intermed wh sup coll
IPF interpeduncular fossa

IPR interpedunc nu, rostral
LEnt lat entorhinal cortex
LMol lacunosum molec lr
MA3 med acc oculomot nu
MGD med genic nu, dors pt
MGM med genic nu, med

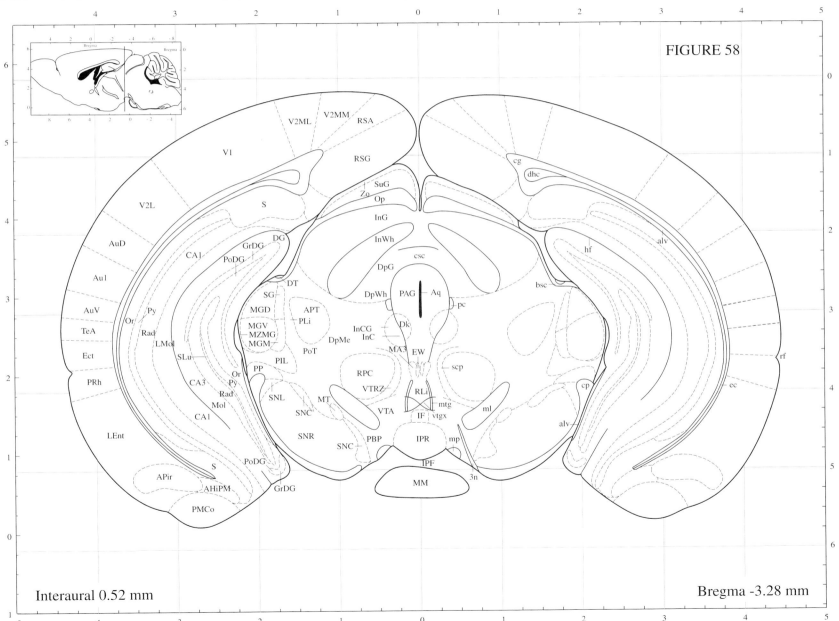

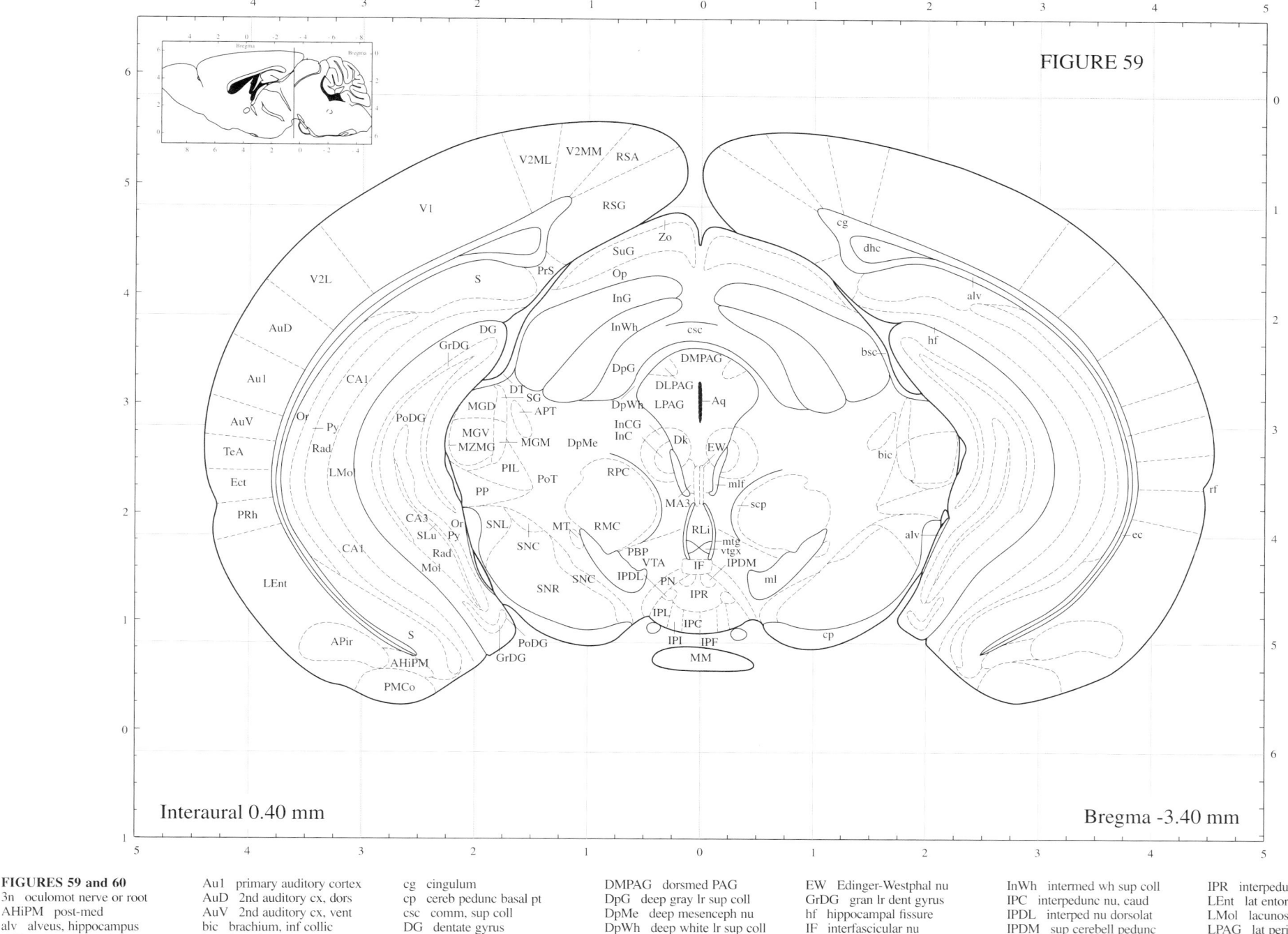

FIGURE 59

Interaural 0.40 mm

Bregma -3.40 mm

FIGURES 59 and 60

3n oculomot nerve or root
AHiPM post-med
alv alveus, hippocampus
APir amygpir transit area
APT ant pretectal nu
Aq aqueduct (Sylvius)

Au1 primary auditory cortex
AuD 2nd auditory cx, dors
AuV 2nd auditory cx, vent
bic brachium, inf collic
bsc brachium sup colliculus
CA1 field CA1, hipp
CA3 field CA3, hipp

cg cingulum
cp cereb pedunc basal pt
csc comm, sup coll
DG dentate gyrus
dhc dors hipp comm
Dk nu, Darkschewitsch
DLPAG dorsolat PAG

DMPAG dorsmed PAG
DpG deep gray lr sup coll
DpMe deep mesenceph nu
DpWh deep white lr sup coll
DT dors term nu acc opt tr
ec ext capsule
Ect ectorhinal cortex

EW Edinger-Westphal nu
GrDG gran lr dent gyrus
hf hippocampal fissure
IF interfascicular nu
InC interstit nu Cajal
InCG inters nu Cajal, great
InG intermed gray lr sup coll

InWh intermed wh sup coll
IPC interpedunc nu, caud
IPDL interped nu dorsolat
IPDM sup cerebell pedunc
IPF interpeduncular fossa
IPI interped nu, intermed
IPL interpedunc nu, lat

IPR interpedunc nu, rostral
LEnt lat entorhinal cortex
LMol lacunosum molec lr
LPAG lat periaqueduct gray
MA3 med acc oculomot nu
MGD med genic nu, dors pt
MGM med genic nu, med

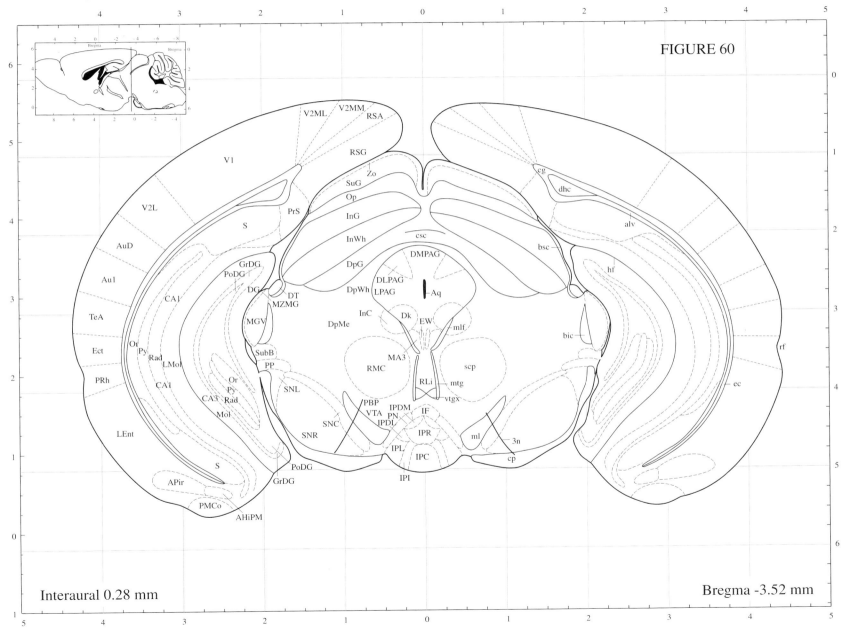

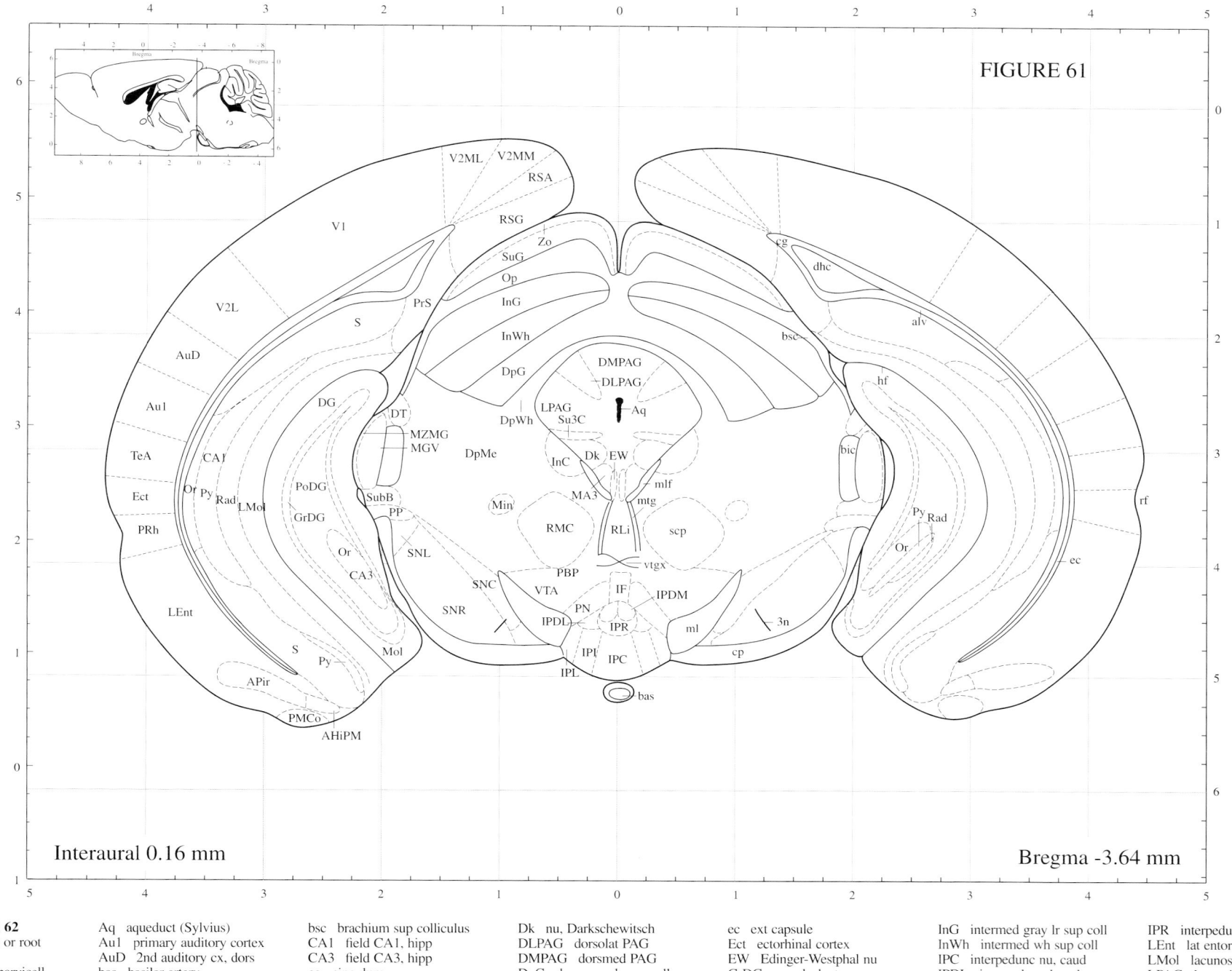

FIGURE 61

Interaural 0.16 mm

Bregma -3.64 mm

FIGURES 61 and 62

3n   oculomot nerve or root
3N   oculomot nu
3PC oculomot nu parvicell
AHiPM post-med
alv alveus, hippocampus
APir amygpir transit area

Aq   aqueduct (Sylvius)
Au1 primary auditory cortex
AuD 2nd auditory cx, dors
bas basilar artery
BIC nu brachium inf collic
bic brachium, inf collic
bp   brachium pontis

bsc brachium sup colliculus
CA1 field CA1, hipp
CA3 field CA3, hipp
cg   cingulum
cp   cereb pedunc basal pt
DG   dentate gyrus
dhc dors hipp comm

Dk   nu, Darkschewitsch
DLPAG dorsolat PAG
DMPAG dorsmed PAG
DpG deep gray lr sup coll
DpMe deep mesenceph nu
DpWh deep white lr sup coll
DT   dors term nu acc opt tr

ec   ext capsule
Ect ectorhinal cortex
EW   Edinger-Westphal nu
GrDG gran lr dent gyrus
hf   hippocampal fissure
IF   interfascicular nu
InC interstit nu Cajal

InG intermed gray lr sup coll
InWh intermed wh sup coll
IPC interpedunc nu, caud
IPDL interped nu dorsolat
IPDM sup cerebell pedunc
IPI interped nu, intermed
IPL interpedunc nu, lat

IPR interpedunc nu, rostral
LEnt lat entorhinal cortex
LMol lacunosum molec lr
LPAG lat periaqueduct gray
MA3 med acc oculomot nu
MGV med genic nu, vent pt
Min minimus nu

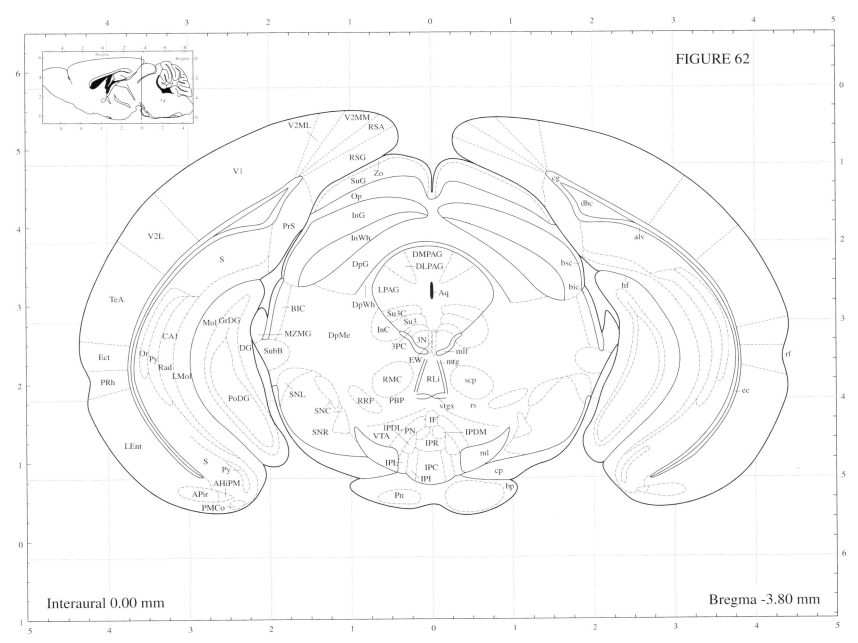

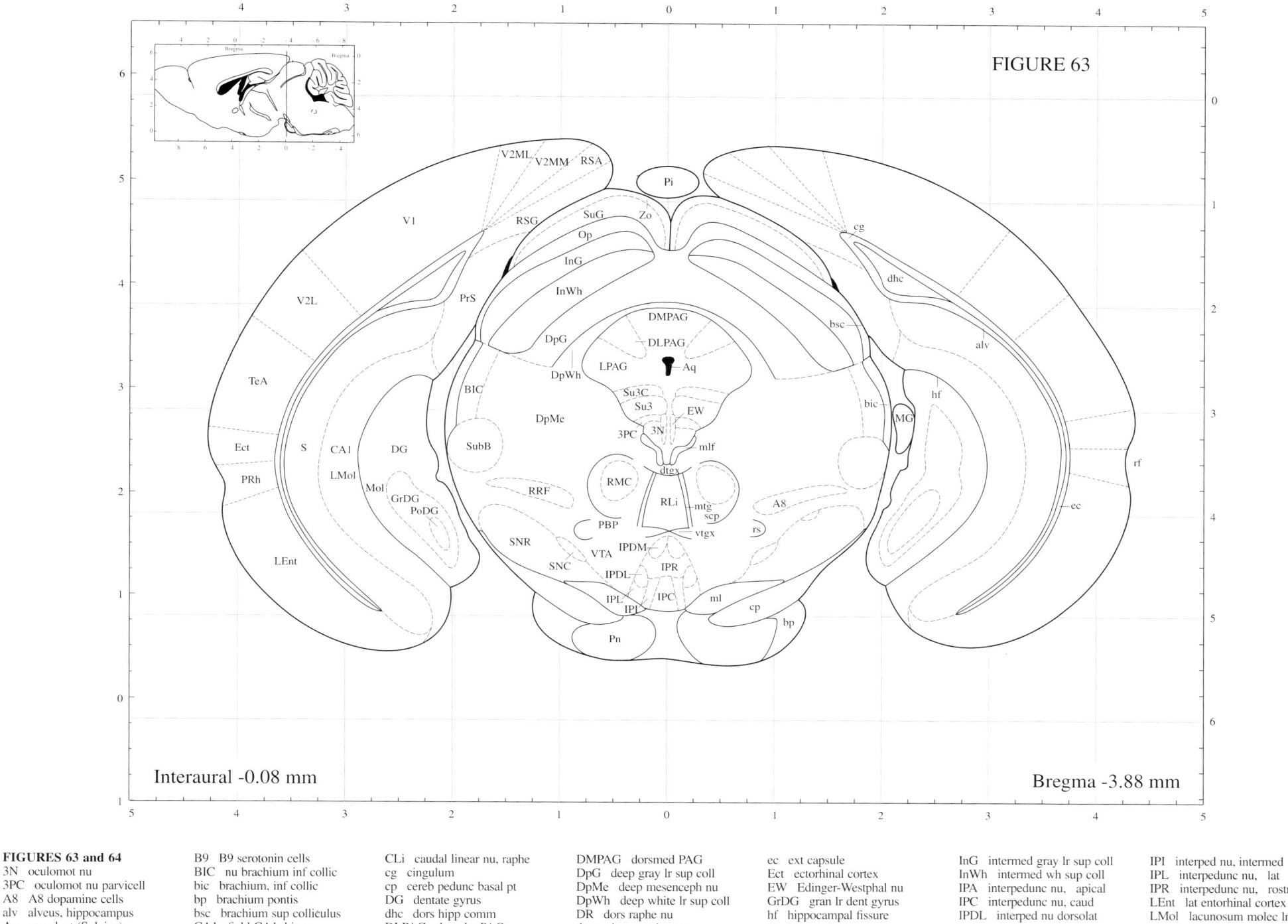

FIGURE 63

Interaural -0.08 mm

Bregma -3.88 mm

FIGURES 63 and 64

3N   oculomot nu
3PC   oculomot nu parvicell
A8   A8 dopamine cells
alv   alveus, hippocampus
Aq   aqueduct (Sylvius)

B9   B9 serotonin cells
BIC   nu brachium inf collic
bic   brachium, inf collic
bp   brachium pontis
bsc   brachium sup colliculus
CA1   field CA1, hipp

CLi   caudal linear nu, raphe
cg   cingulum
cp   cereb pedunc basal pt
DG   dentate gyrus
dhc   dors hipp comm
DLPAG   dorsolat PAG

DMPAG   dorsmed PAG
DpG   deep gray lr sup coll
DpMe   deep mesenceph nu
DpWh   deep white lr sup coll
DR   dors raphe nu
dtgx   dors teg decuss

ec   ext capsule
Ect   ectorhinal cortex
EW   Edinger-Westphal nu
GrDG   gran lr dent gyrus
hf   hippocampal fissure
I3   interoculomot nu

InG   intermed gray lr sup coll
InWh   intermed wh sup coll
IPA   interpedunc nu, apical
IPC   interpedunc nu, caud
IPDL   interped nu dorsolat
IPDM   sup cerebell pedunc

IPI   interped nu, intermed
IPL   interpedunc nu, lat
IPR   interpedunc nu, rostral
LEnt   lat entorhinal cortex
LMol   lacunosum molec lr
LPAG   lat periaqueduct gray

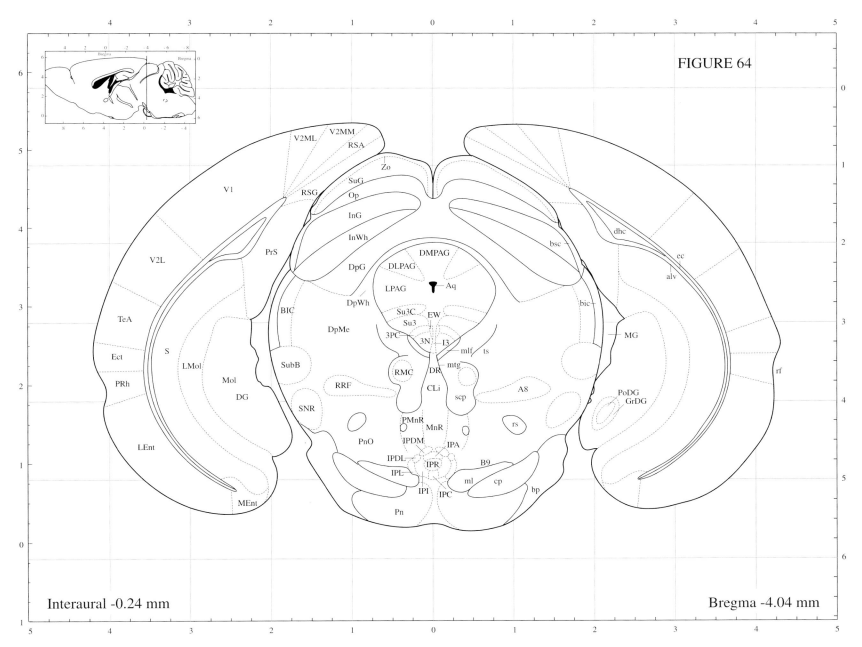

FIGURE 64

Interaural -0.24 mm Bregma -4.04 mm

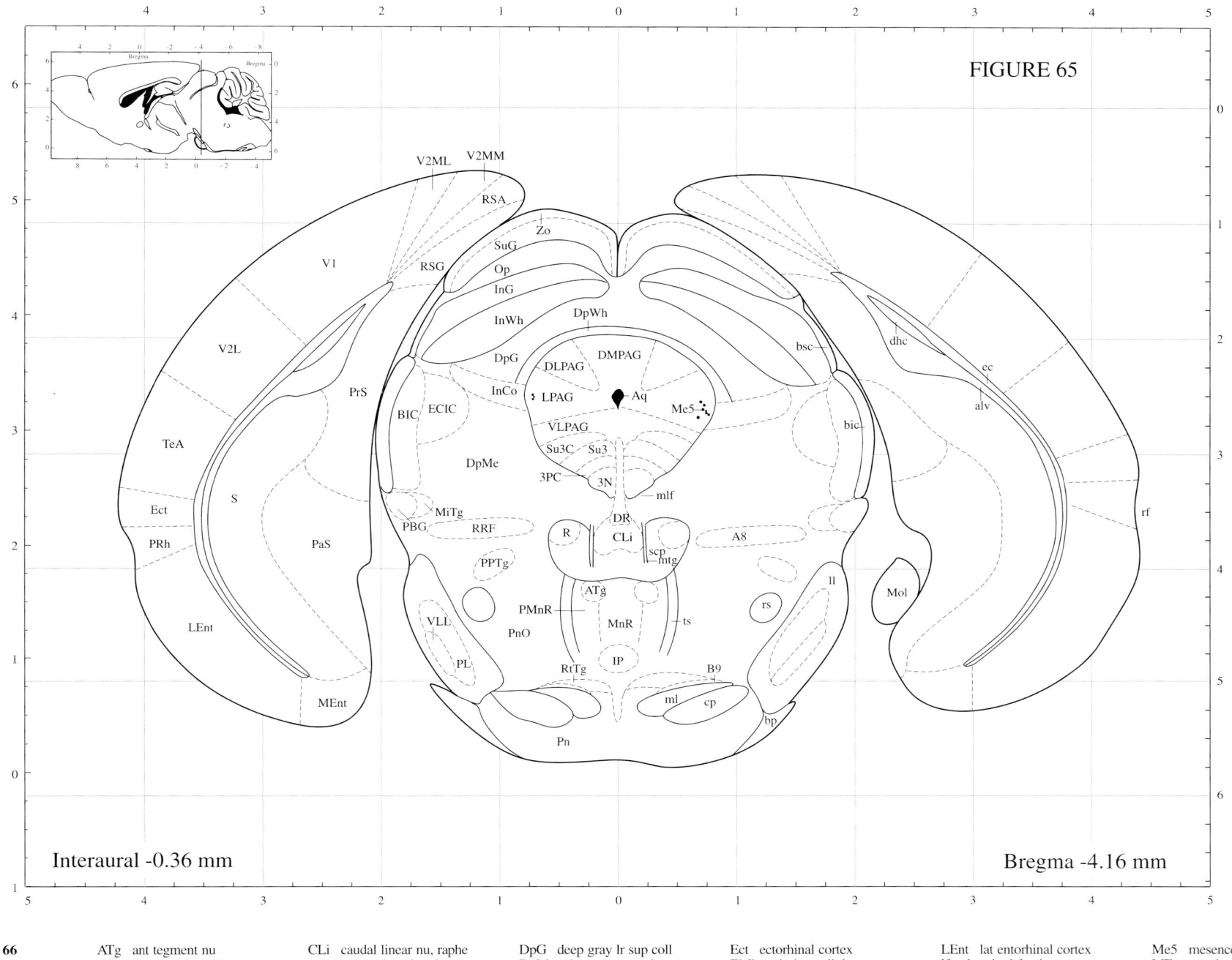

FIGURE 65

Interaural -0.36 mm

Bregma -4.16 mm

FIGURES 65 and 66

3N oculomot nu	ATg ant tegment nu	CLi caudal linear nu, raphe	DpG deep gray lr sup coll	Ect ectorhinal cortex	LEnt lat entorhinal cortex	Me5 mesenceph 5 nu
3PC oculomot nu parvicell	B9 B9 serotonin cells	cp cereb pedunc basal pt	DpMe deep mesenceph nu	EMi epimicrocellular nu	lfp longitud fascic, pons	MEnt med entorhinal cortex
A8 A8 dopamine cells	BIC nu brachium inf collic	dhc dors hipp comm	DpWh deep white lr sup coll	InCo intercollicular nu	ll lat lemniscus	MiTg microcell tegment nu
alv alveus, hippocampus	bic brachium, inf collic	DLPAG dorsolat PAG	DR dors raphe nu	InG intermed gray lr sup coll	LPAG lat periaqueduct gray	ml med lemniscus
Aq aqueduct (Sylvius)	bp brachium pontis	DMPAG dorsmed PAG	ec ext capsule	InWh intermed wh sup coll	m5 motor root, 5 n	mlf med longitud fascic
	bsc brachium sup colliculus	DMPn dorsmed pontine nu	ECIC ext cx, inferior collic	IP interpeduncular nu	mcp middle cerebell pedunc	MnR median raphe nu

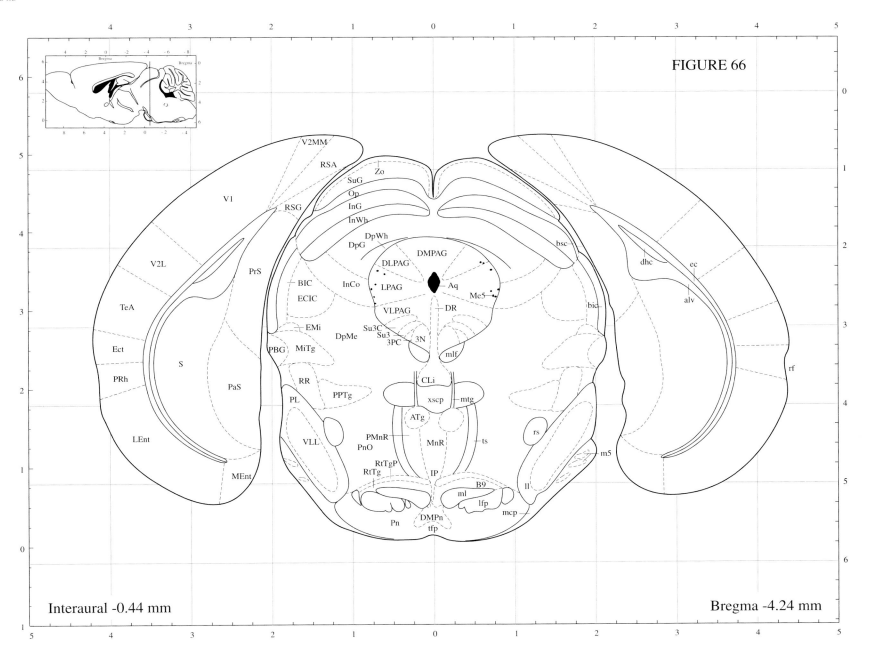

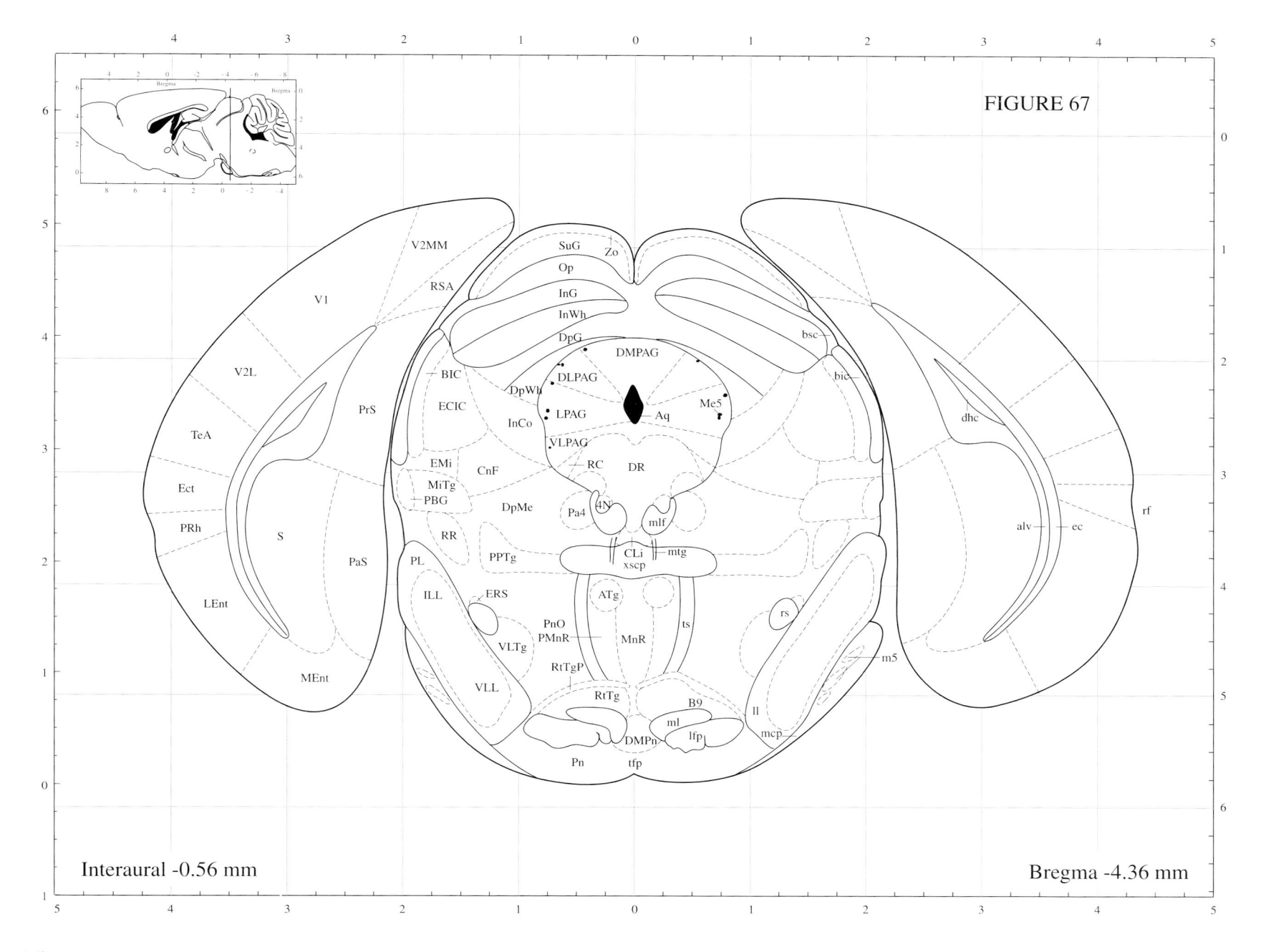

FIGURE 67

Interaural -0.56 mm

Bregma -4.36 mm

FIGURES 67 and 68

4N   trochlear nu
alv   alveus, hippocampus
Aq   aqueduct (Sylvius)
ATg   ant tegment nu
B9   B9 serotonin cells

BIC   nu brachium inf collic
bic   brachium, inf collic
CLi   caudal linear nu, raphe
CnF   cuneiform nu
dhc   dors hipp comm
DLPAG   dorsolat PAG

DMPAG   dorsmed PAG
DMPn   dorsmed pontine nu
DpG   deep gray lr sup coll
DpWh   deep white lr sup coll
DRD   dors raphe nu, dors pt
DRI   dors raphe nu, interfasc

DRV   dors raphe nu, vent pt
DRVL   d raphe nu, ventlat
ECIC   ext cx, inferior collic
ec   ext capsule
Ect   ectorhinal cortex
EMi   epimicrocellular nu

ILL   intermed nu, lat lemnisc
InCo   intercollicular nu
InG   intermed gray lr sup coll
InWh   intermed wh sup coll
LEnt   lat entorhinal cortex
lfp   longitud fascic, pons

ll   lat lemniscus
LPAG   lat periaqueduct gray
m5   motor root, 5 n
mcp   middle cerebell pedunc
Me5   mesenceph 5 nu
MEnt   med entorhinal cortex

MiTg   microcelll tegment nu
ml   med lemniscus
mlf   med longitud fascic
MnR   median raphe nu
mtg   mammillotegment tr
Op   optic n lr sup coll

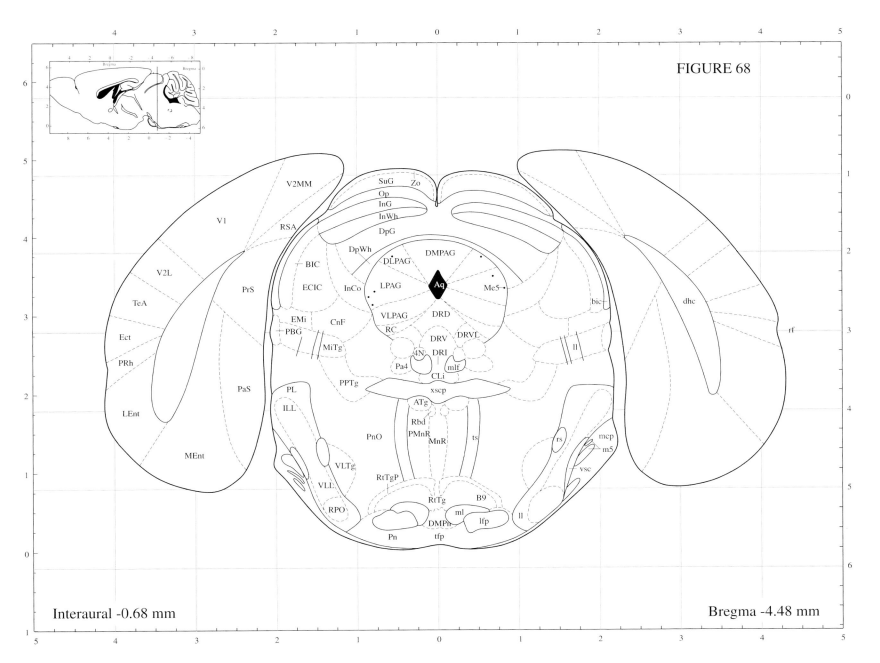

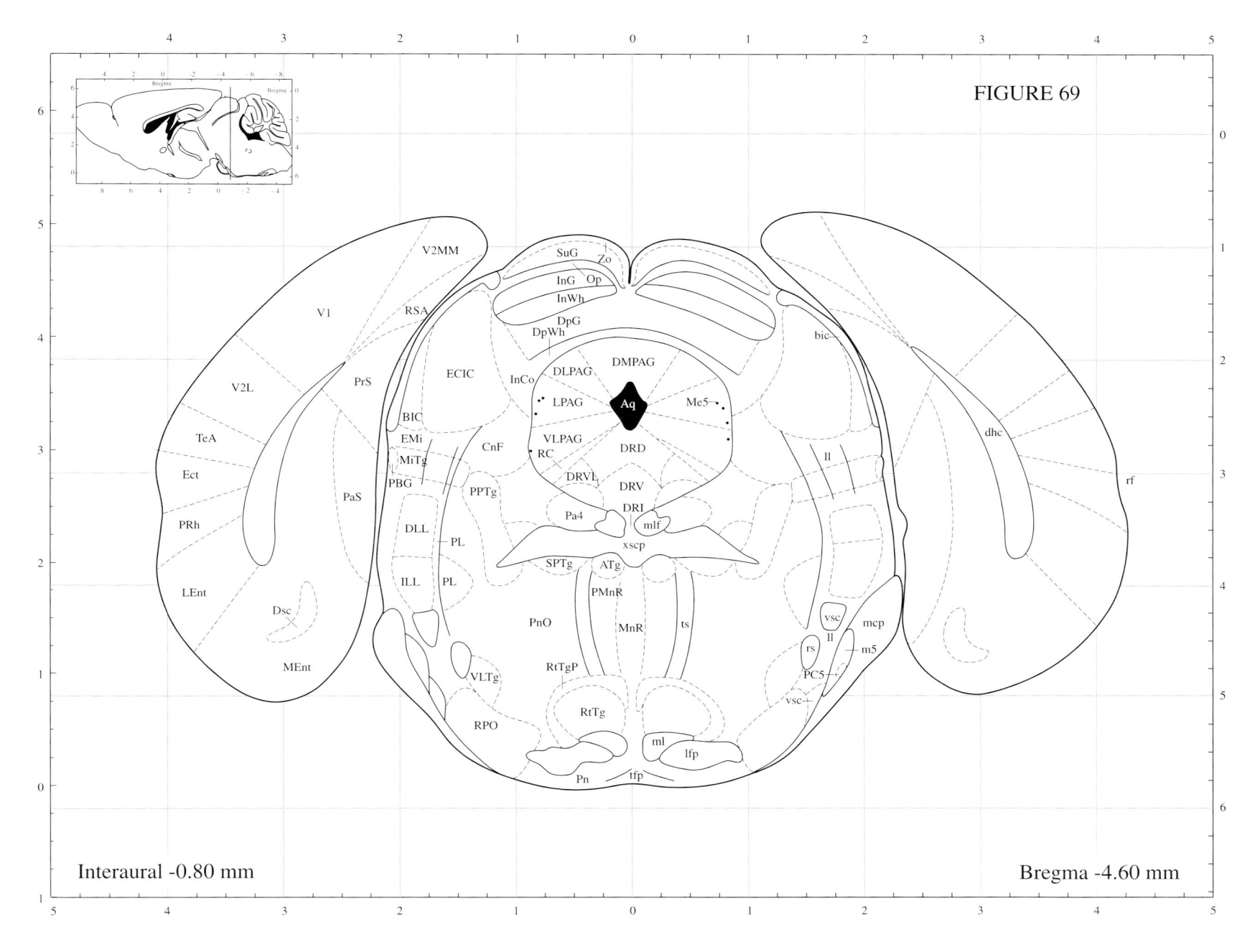

FIGURE 69

Interaural -0.80 mm

Bregma -4.60 mm

FIGURES 69 and 70

l, 2, 3, 4   layer 1, 2, 3, 4
4n   trochlear nerve or root
Aq   aqueduct (Sylvius)
ATg   ant tegment nu
BIC   nu brachium inf collic

bic   brachium, inf collic
CnF   cuneiform nu
dhc   dors hipp comm
DLL   dors nu, lat lemnisc
DLPAG   dorsolat PAG
DMPAG   dorsmed PAG

DpG   deep gray lr sup coll
DpWh   deep white lr sup coll
DRD   dors raphe nu, dors pt
DRI   dors raphe nu, interfasc
DRV   dors raphe nu, vent pt
DRVL   d raphe nu, ventlat

Dsc   lamina diss entorhin cx
ECIC   ext cx, inferior collic
Ect   ectorhinal cortex
EMi   epimicrocellular nu
ILL   intermed nu, lat lemnisc
InCo   intercollicular nu

InG   intermed gray lr sup coll
InWh   intermed wh sup coll
LEnt   lat entorhinal cortex
lfp   longitud fascic, pons
ll   lat lemniscus
LPAG   lat periaqueduct gray

LVPO   laterovent perioliv
m5   motor root, 5 n
mcp   middle cerebell pedunc
Me5   mesenceph 5 nu
MEnt   med entorhinal cortex
MiTg   microcelll tegment nu

ml   med lemniscus
mlf   med longitud fascic
MnR   median raphe nu
MVPO   mediovent perioliv
Op   optic n lr sup coll
P5   peritrigeminal zone

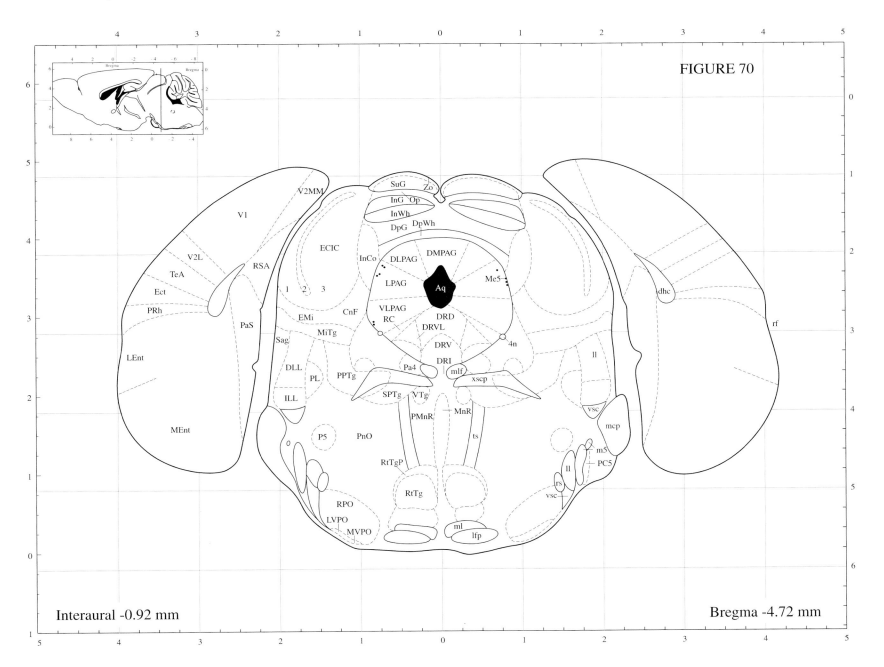

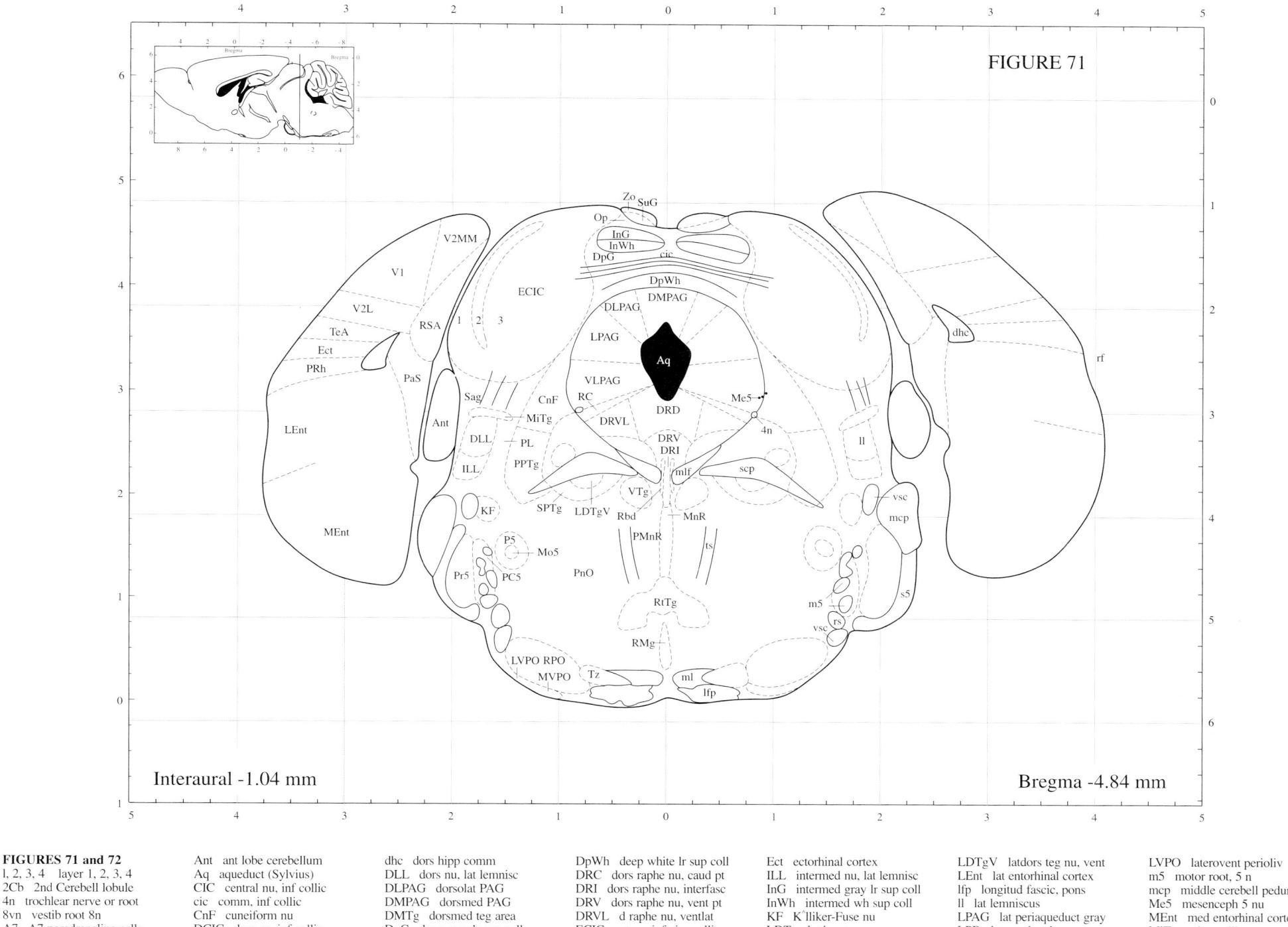

FIGURE 71

Interaural -1.04 mm

Bregma -4.84 mm

FIGURES 71 and 72

l, 2, 3, 4   layer 1, 2, 3, 4
2Cb   2nd Cerebell lobule
4n   trochlear nerve or root
8vn   vestib root 8n
A7   A7 noradrenaline cells

Ant   ant lobe cerebellum
Aq   aqueduct (Sylvius)
CIC   central nu, inf collic
cic   comm, inf collic
CnF   cuneiform nu
DCIC   dors cx, inf collic

dhc   dors hipp comm
DLL   dors nu, lat lemnisc
DLPAG   dorsolat PAG
DMPAG   dorsmed PAG
DMTg   dorsmed teg area
DpG   deep gray lr sup coll

DpWh   deep white lr sup coll
DRC   dors raphe nu, caud pt
DRI   dors raphe nu, interfasc
DRV   dors raphe nu, vent pt
DRVL   d raphe nu, ventlat
ECIC   ext cx, inferior collic

Ect   ectorhinal cortex
ILL   intermed nu, lat lemnisc
InG   intermed gray lr sup coll
InWh   intermed wh sup coll
KF   Kölliker-Fuse nu
LDTg   latdors teg nu

LDTgV   latdors teg nu, vent
LEnt   lat entorhinal cortex
lfp   longitud fascic, pons
ll   lat lemniscus
LPAG   lat periaqueduct gray
LPB   lat parabrach nu

LVPO   laterovent perioliv
m5   motor root, 5 n
mcp   middle cerebell pedunc
Me5   mesenceph 5 nu
MEnt   med entorhinal cortex
MiTg   microcelll tegment nu

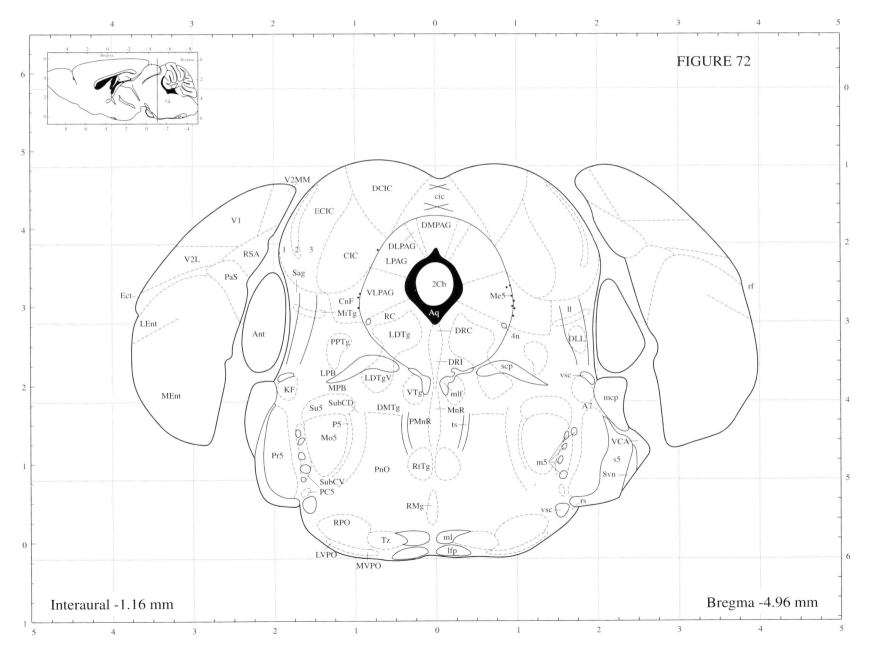

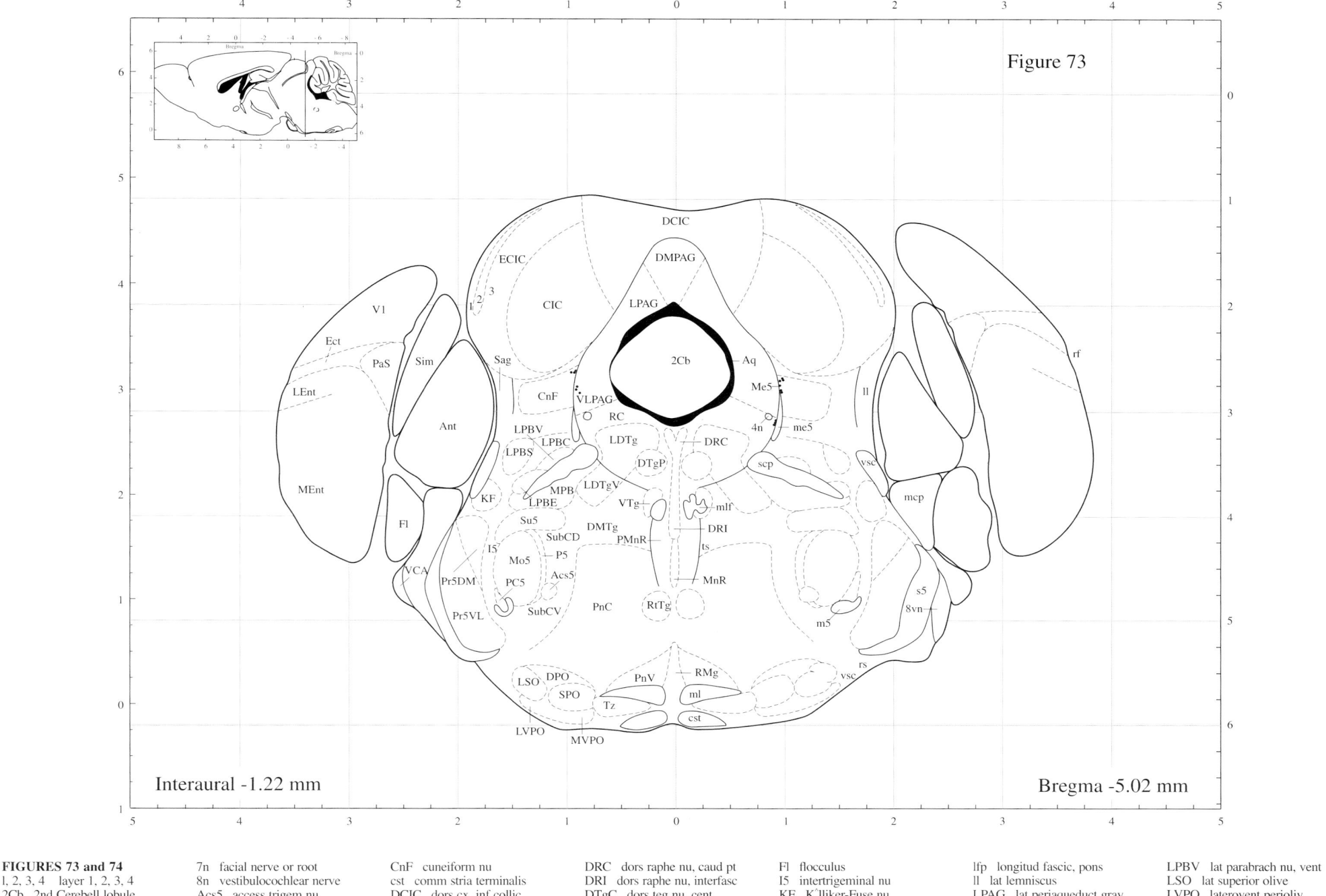

Figure 73

Interaural -1.22 mm

Bregma -5.02 mm

FIGURES 73 and 74

l, 2, 3, 4 layer 1, 2, 3, 4	7n facial nerve or root	CnF cuneiform nu	DRC dors raphe nu, caud pt	Fl flocculus	lfp longitud fascic, pons	LPBV lat parabrach nu, vent
2Cb 2nd Cerebell lobule	8n vestibulocochlear nerve	cst comm stria terminalis	DRI dors raphe nu, interfasc	I5 intertrigeminal nu	ll lat lemniscus	LSO lat superior olive
3Cb 3rd Cerebell lobule	Acs5 access trigem nu	DCIC dors cx, inf collic	DTgC dors teg nu, cent	KF K´lliker-Fuse nu	LPAG lat periaqueduct gray	LVPO laterovent perioliv
4n trochlear nerve or root	Ant ant lobe cerebellum	DMPAG dorsmed PAG	DTgP dors teg nu, pericent	LDTg latdors teg nu	LPBC lat parabrach nu, ce	m5 motor root, 5 n
4V 4th ventric	Aq aqueduct (Sylvius)	DMTg dorsmed teg area	ECIC ext cx, inferior collic	LDTgV latdors teg nu, vent	LPBD lat parabrach nu dors	mcp middle cerebell pedunc
	CIC central nu, inf collic	DPO dors perioliv	Ect ectorhinal cortex	LEnt lat entorhinal cortex	LPBE lat parabrach nu, ext	Me5 mesenceph 5 nu

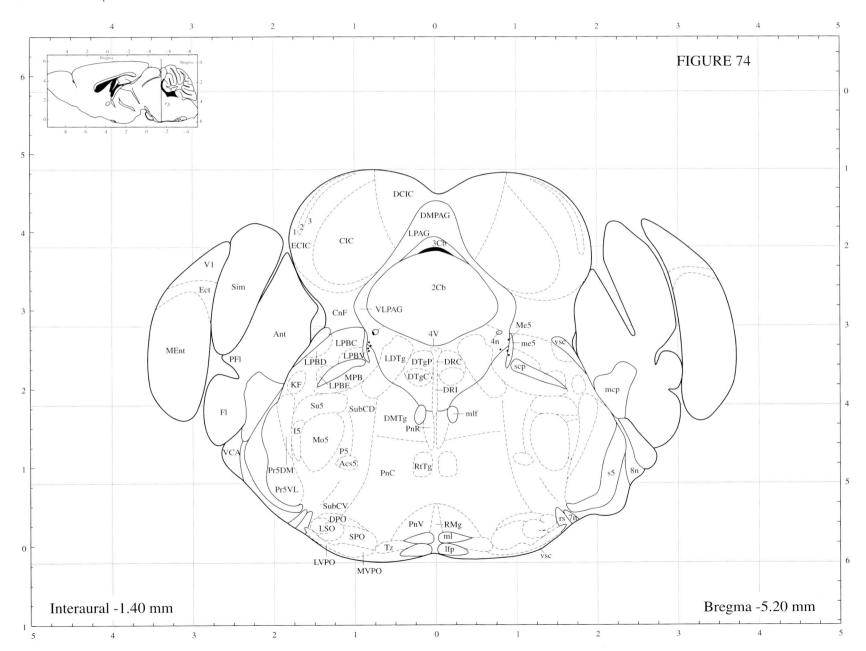

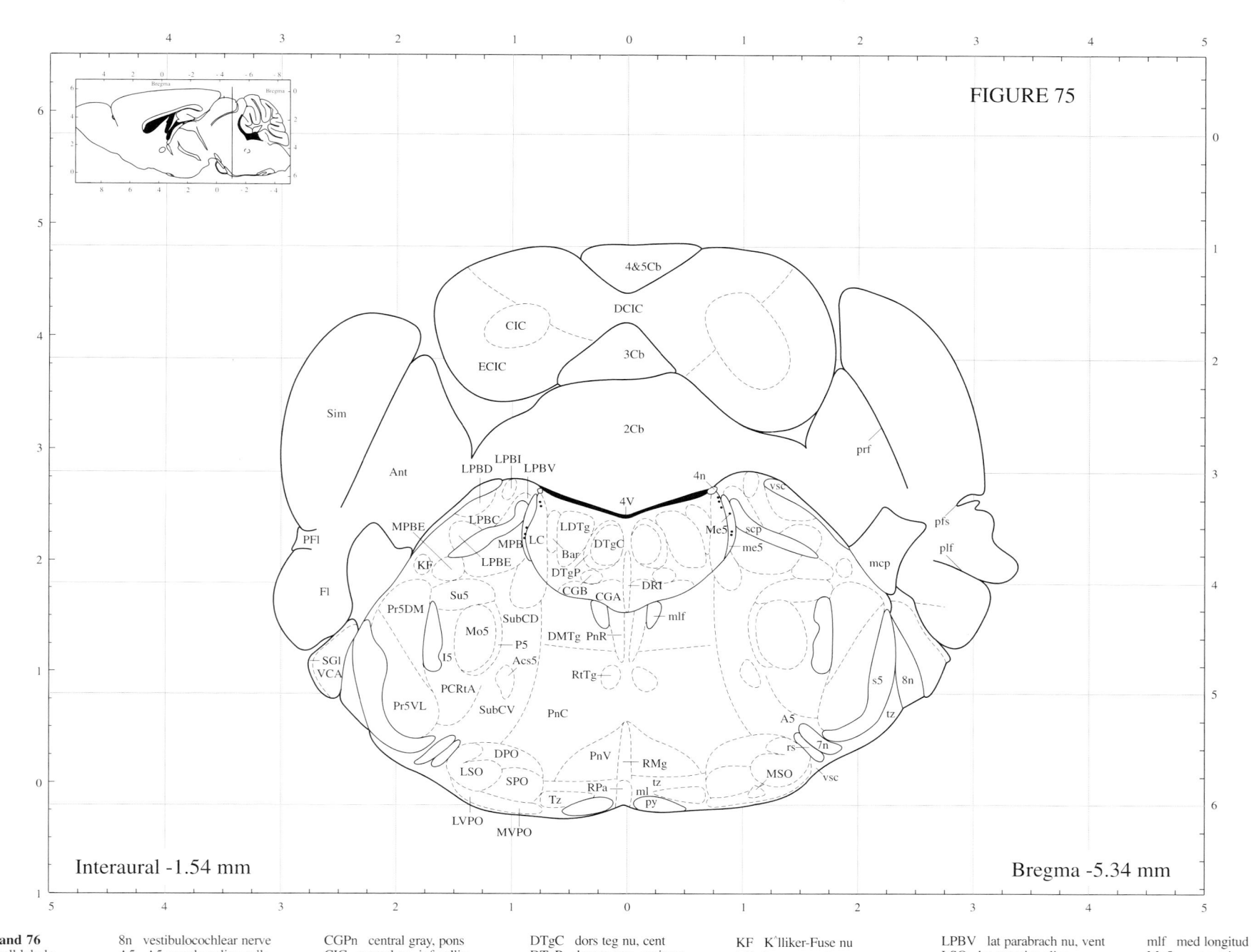

FIGURE 75

Interaural -1.54 mm

Bregma -5.34 mm

FIGURES 75 and 76

2Cb 2nd Cerebell lobule
3Cb 3rd Cerebell lobule
4&5Cb 4&5th Cerebell lob
4n trochlear nerve or root
4V 4th ventric
7n facial nerve or root

8n vestibulocochlear nerve
A5 A5 noradrenaline cells
Acs5 access trigem nu
Ant ant lobe cerebellum
Bar Barrington's nu
CGA central gray, · pt
CGB central gray, beta pt

CGPn central gray, pons
CIC central nu, inf collic
Crus1 crus 1, ansiform lob
DCIC dors cx, inf collic
DMTg dorsmed teg area
DPO dors perioliv
DRI dors raphe nu, interfasc

DTgC dors teg nu, cent
DTgP dors teg nu, pericent
ECIC ext cx, inferior collic
Fl flocculus
GiA gigantocell retic nu, ·
I5 intertrigeminal nu
IRt intermed retic nu

KF K'lliker-Fuse nu
LC locus coeruleus
LDTg latdors teg nu
LPBC lat parabrach nu, ce
LPBD lat parabrach nu dors
LPBE lat parabrach nu, ext
LPBI lat parabrach nu, int

LPBV lat parabrach nu, vent
LSO lat superior olive
LVPO laterovent perioliv
mcp middle cerebell pedunc
Me5 mesenceph 5 nu
me5 mesenceph 5 tr
ml med lemniscus

mlf med longitud fascic
Mo5 motor trigeminal nu
MPB med parabrach nu
MPBE med parabrach nu ext
MSO med superior olive
MVPO mediovent perioliv
O nu O

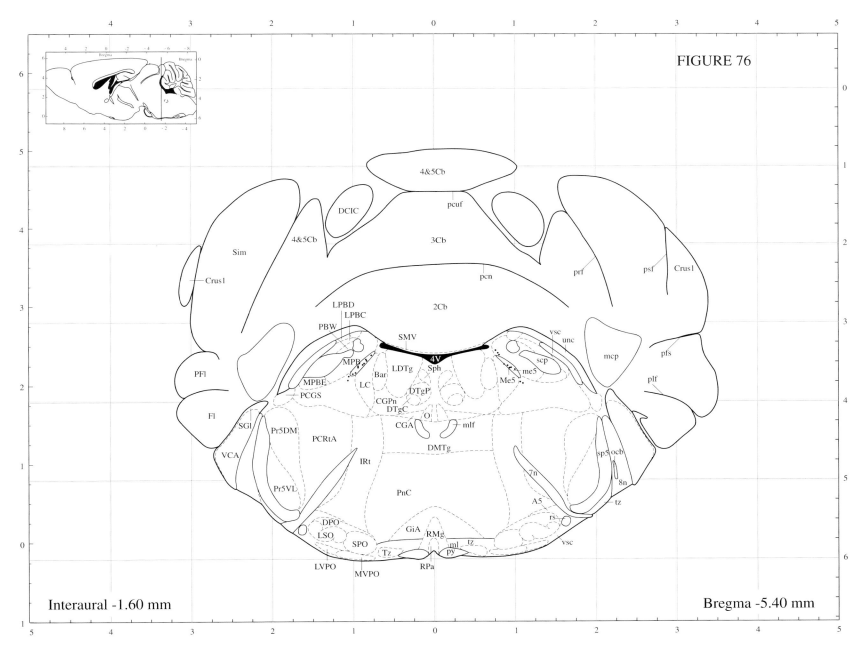

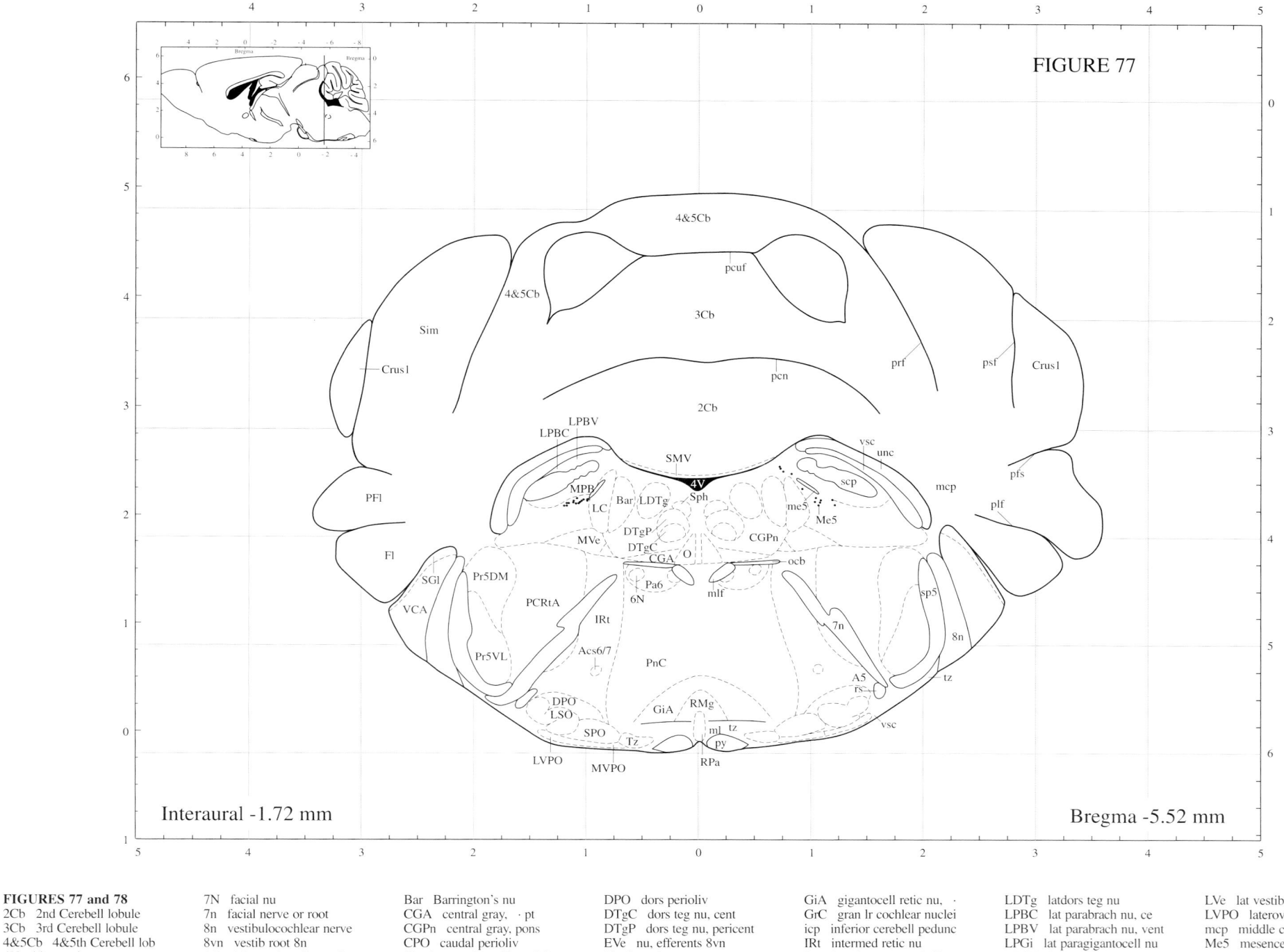

FIGURE 77

Interaural -1.72 mm

Bregma -5.52 mm

FIGURES 77 and 78
2Cb 2nd Cerebell lobule
3Cb 3rd Cerebell lobule
4&5Cb 4&5th Cerebell lob
4V 4th ventric
6N abducens nu

7N facial nu
7n facial nerve or root
8n vestibulocochlear nerve
8vn vestib root 8n
A5 A5 noradrenaline cells
Acs7 accessory facial nu

Bar Barrington's nu
CGA central gray, · pt
CGPn central gray, pons
CPO caudal perioliv
Crus1 crus 1, ansiform lob
DC dors cochlear nu

DPO dors perioliv
DTgC dors teg nu, cent
DTgP dors teg nu, pericent
EVe nu, efferents 8vn
Fl flocculus
Gi gigantocell retic nu

GiA gigantocell retic nu, ·
GrC gran lr cochlear nuclei
icp inferior cerebell pedunc
IRt intermed retic nu
Lat lat (dentate) cerebell nu
LC locus coeruleus

LDTg latdors teg nu
LPBC lat parabrach nu, ce
LPBV lat parabrach nu, vent
LPGi lat paragigantocell nu
LR4V lat recess, 4th ventric
LSO lat superior olive

LVe lat vestibular nu
LVPO laterovent perioliv
mcp middle cerebell pedunc
Me5 mesenceph 5 nu
me5 mesenceph 5 tr
ml med lemniscus

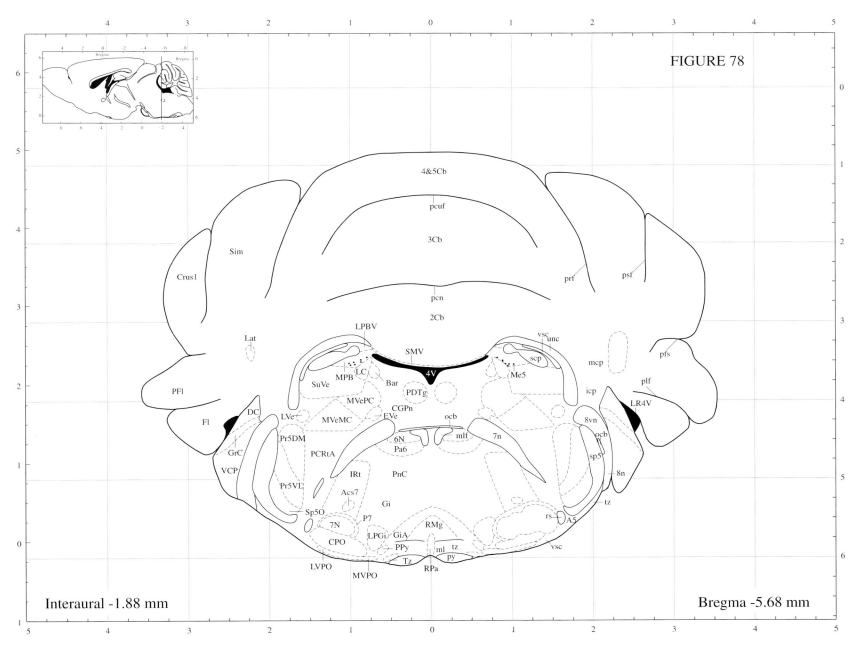

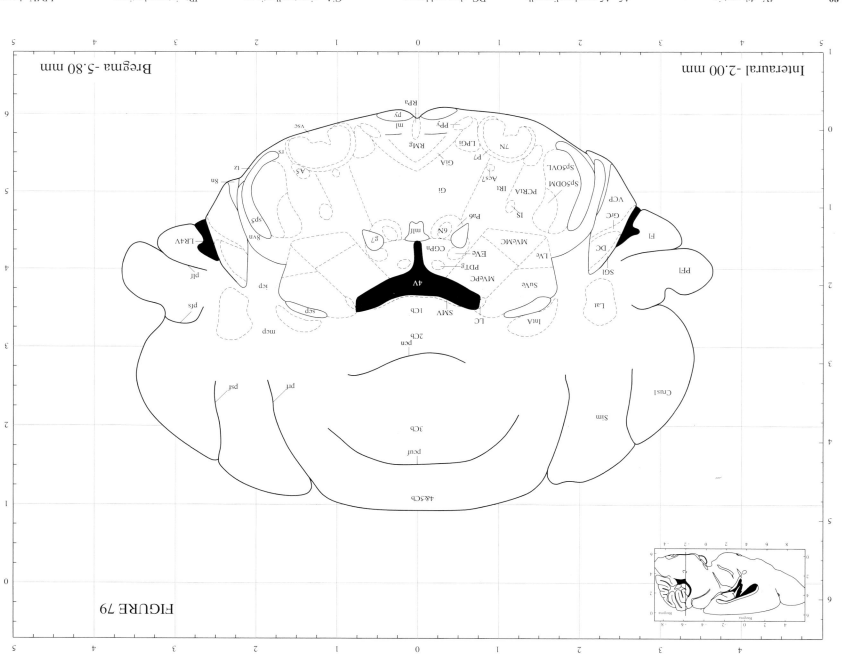

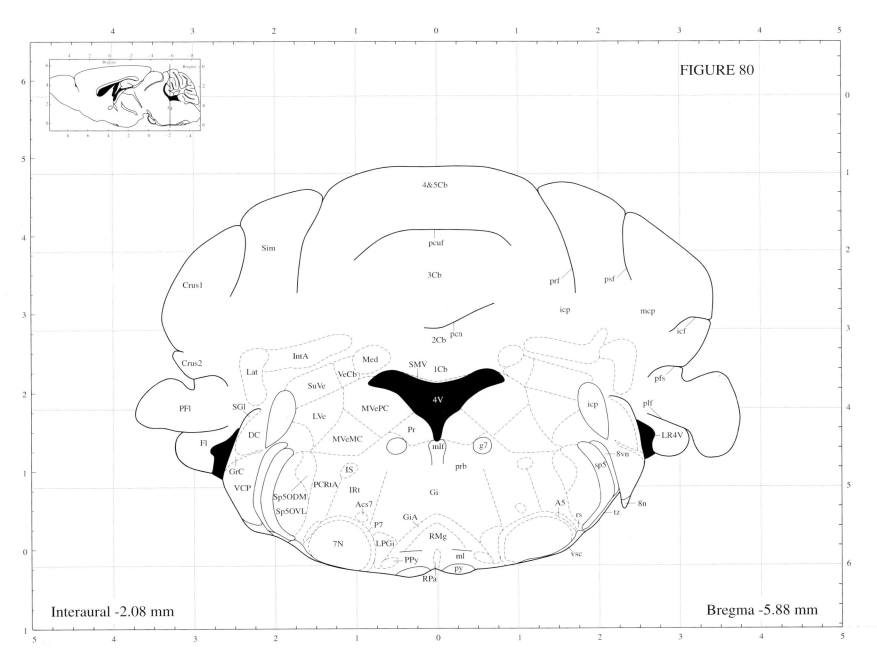

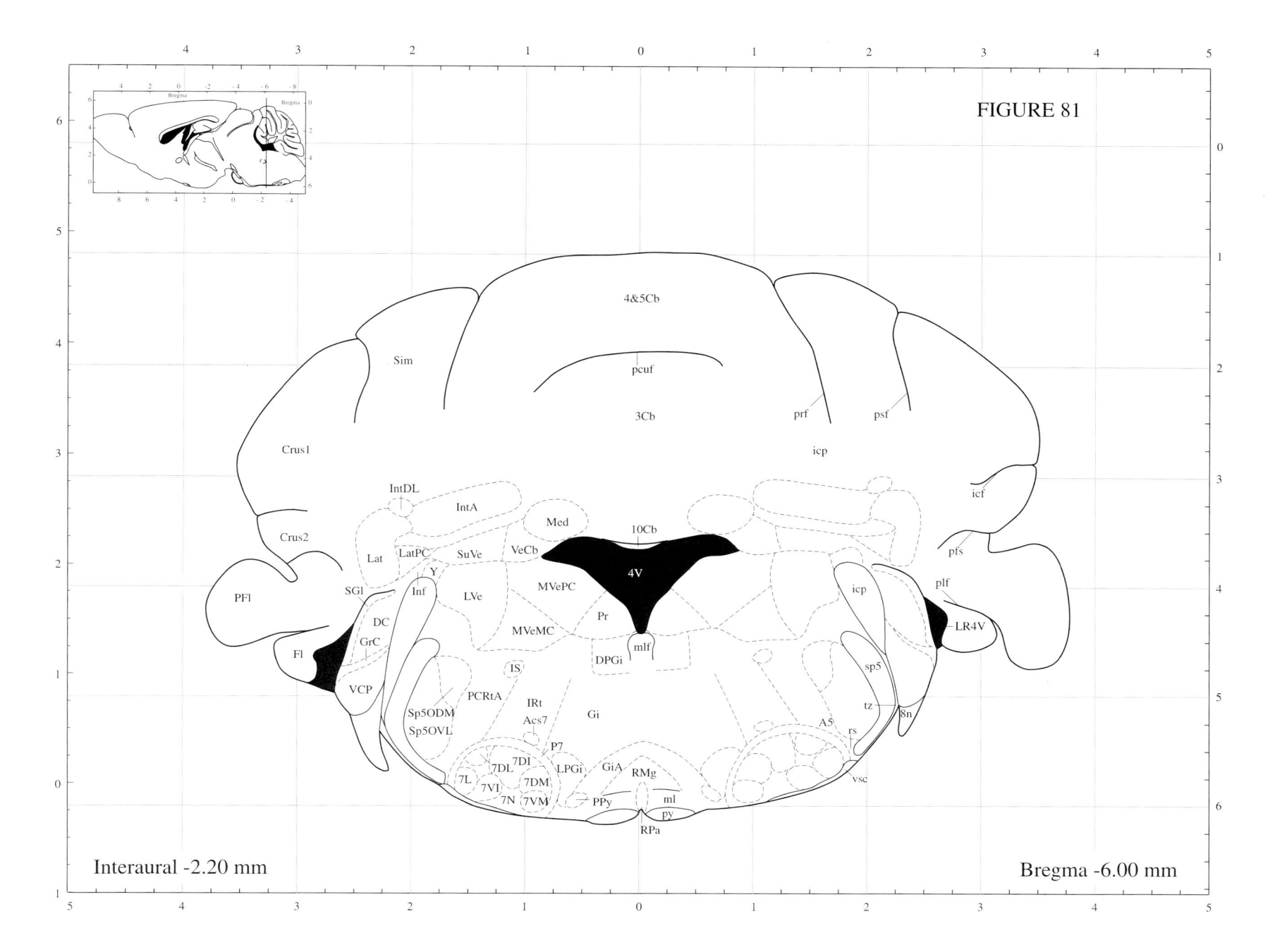

FIGURE 81

Interaural -2.20 mm

Bregma -6.00 mm

FIGURES 81 and 82

3Cb 3rd Cerebell lobule
4&5Cb 4&5th Cerebell lob
4V 4th ventric
7DI facial nu dors intermed

7DL facial nu, dorsolat
7DM facial nu, dorsmed
7L facial nu, lat
7N facial nu
7VI facial nu, vent intermed

7VM facial nu, ventmed
8n vestibulocochlear nerve
A5 A5 noradrenaline cells
10Cb 10th Cerebell lobule
Acs7 accessory facial nu

Crus1 crus 1, ansiform lob
Crus2 crus 2, ansiform lob
DC dors cochlear nu
DMSp5 dorsmed spin 5 nu
DPGi dors paragigantcell nu

Fl flocculus
Gi gigantocell retic nu
GiA gigantocell retic nu, ·
GrC gran lr cochlear nuclei
icf intercrural fissure

icp inferior cerebell pedunc
Inf infracerebell nu
IntA interpos Cb nu, ant
IntDL interpos Cb nu, dl
IRt intermed retic nu

IS inferior salivatory nu
Lat lat (dentate) cerebell nu
LatPC lat Cb nu, parvicell
LPGi lat paragigantcell nu
LR4V lat recess, 4th ventric

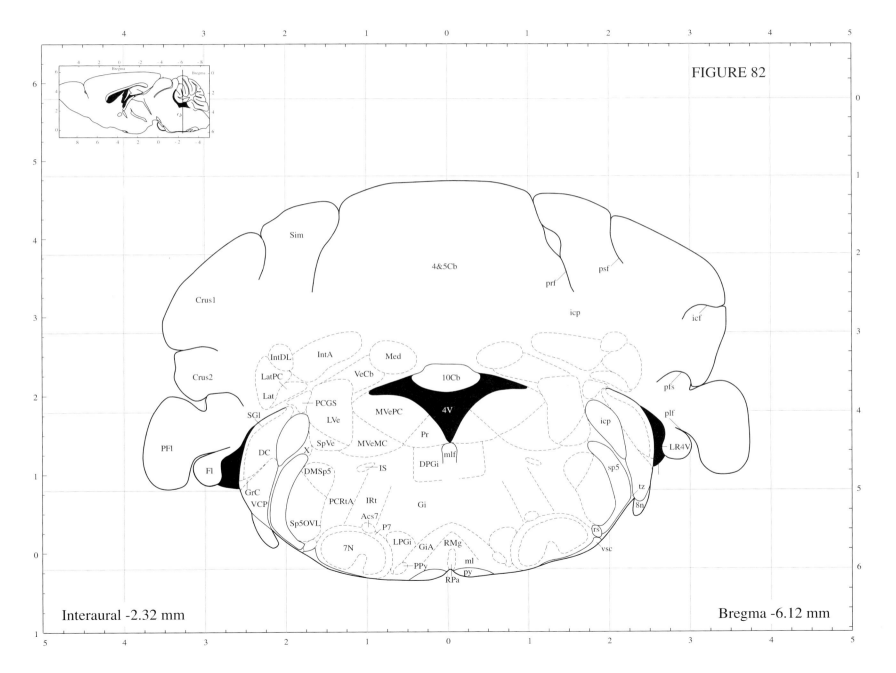

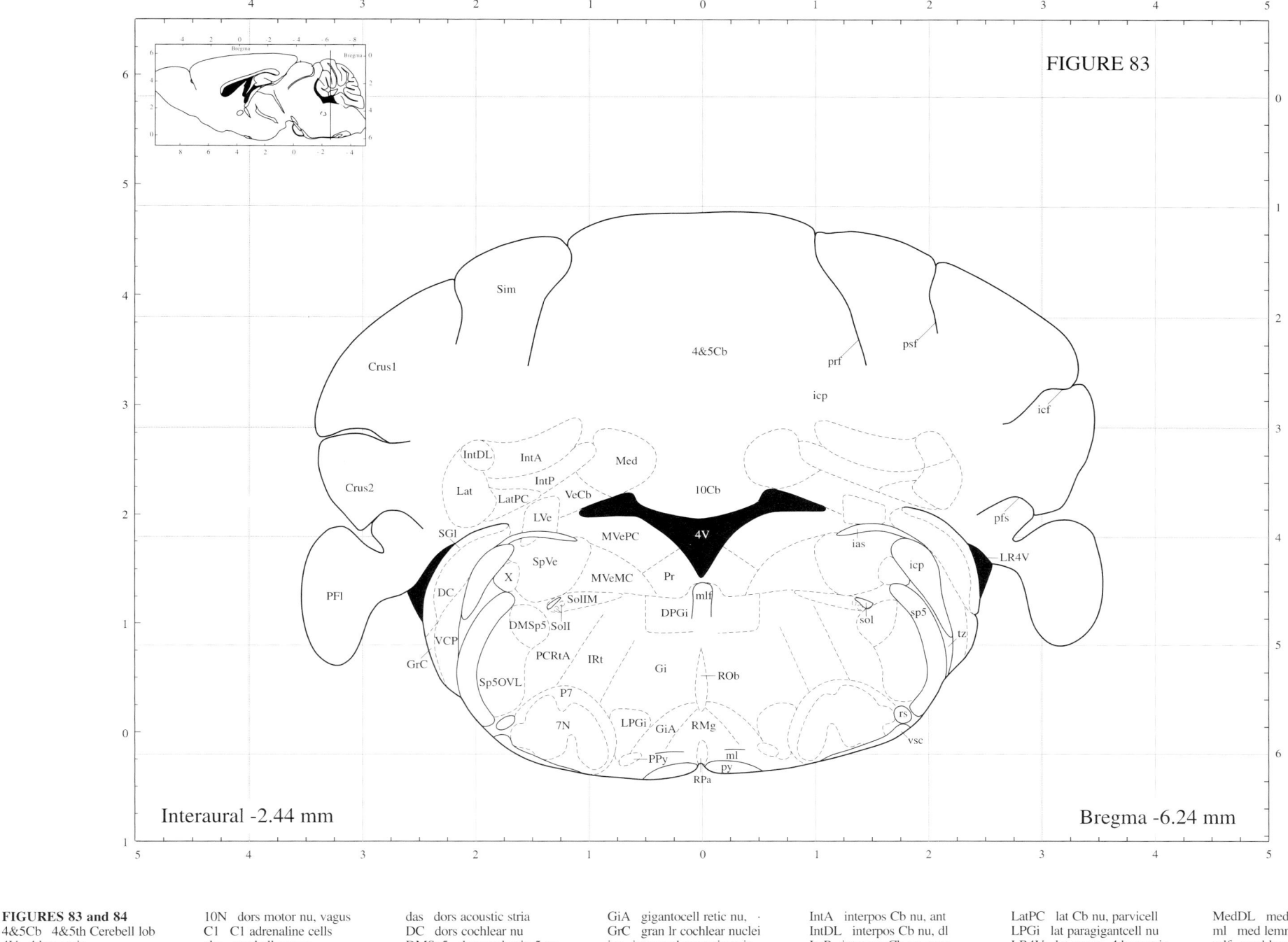

FIGURE 83

Interaural -2.44 mm

Bregma -6.24 mm

FIGURES 83 and 84

4&5Cb 4&5th Cerebell lob	10N dors motor nu, vagus	das dors acoustic stria	GiA gigantocell retic nu, ·	IntA interpos Cb nu, ant	LatPC lat Cb nu, parvicell	MedDL med Cb nu, dl pro
4V 4th ventric	C1 C1 adrenaline cells	DC dors cochlear nu	GrC gran lr cochlear nuclei	IntDL interpos Cb nu, dl	LPGi lat paragigantcell nu	ml med lemniscus
7N facial nu	cbc cerebell comm	DMSp5 dorsmed spin 5 nu	ias intermed acoustic stria	IntP interpos Cb nu, post	LR4V lat recess, 4th ventric	mlf med longitud fascic
10Cb 10th Cerebell lobule	Crus1 crus 1, ansiform lob	DPGi dors paragigantcell nu	icf intercrural fissure	IRt intermed retic nu	LVe lat vestibular nu	MVeMC med vestib, mcell
	Crus2 crus 2, ansiform lob	Gi gigantocell retic nu	icp inferior cerebell pedunc	Lat lat (dentate) cerebell nu	Med med cerebell nu	MVePC med vestib, pcell

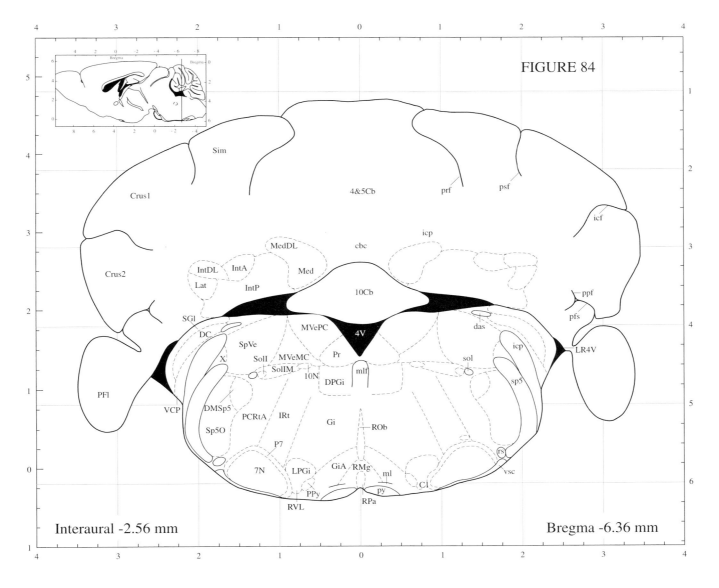

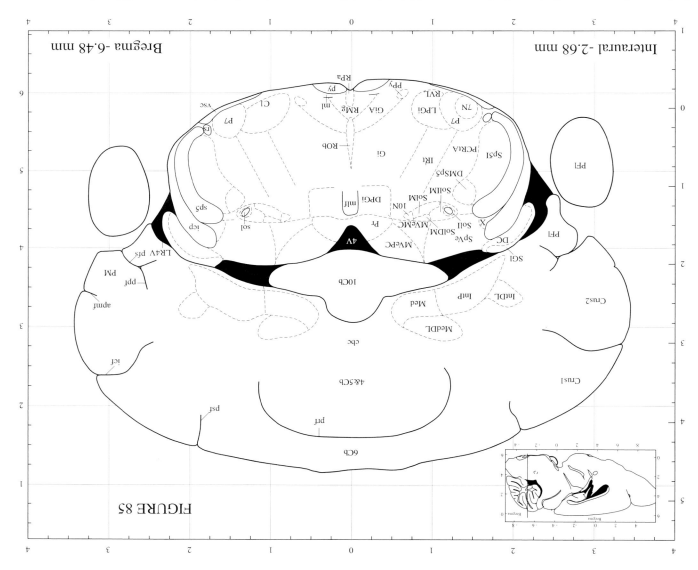

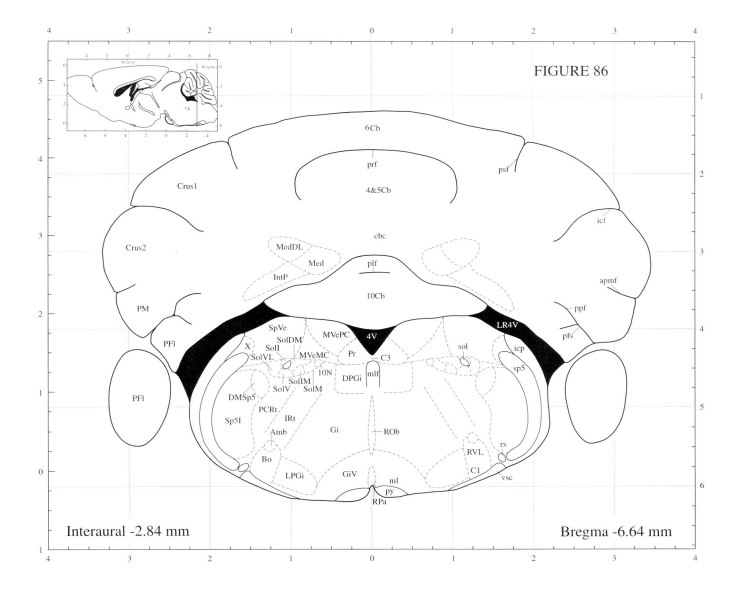

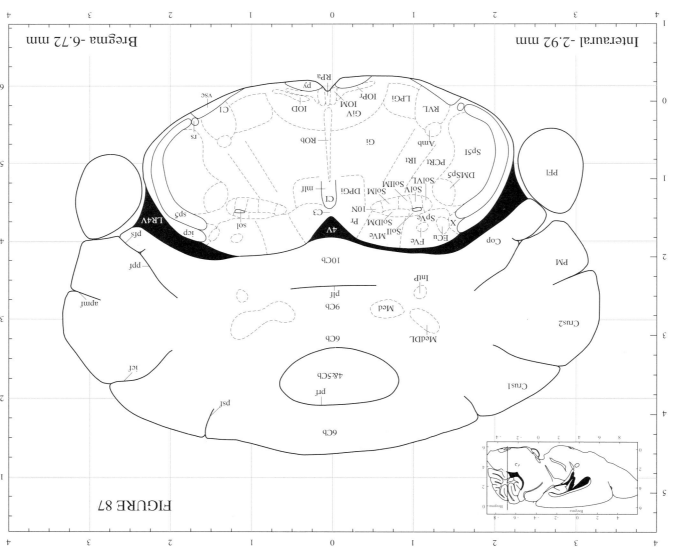

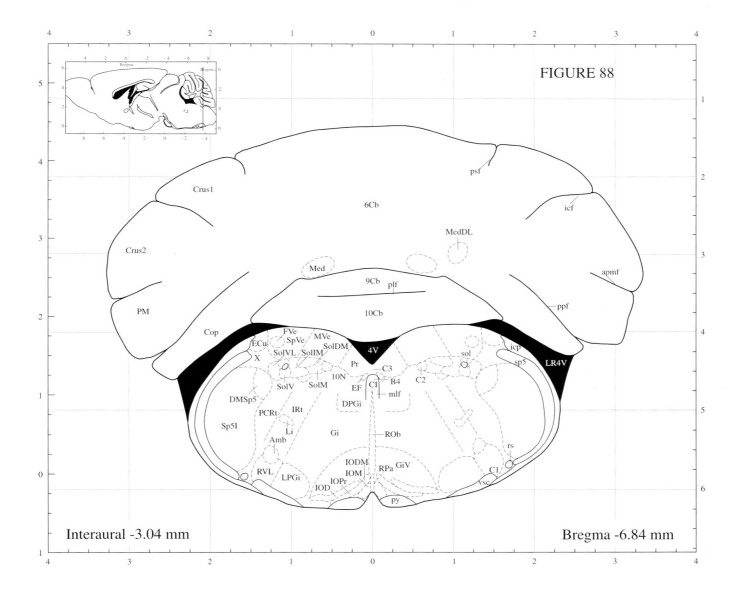

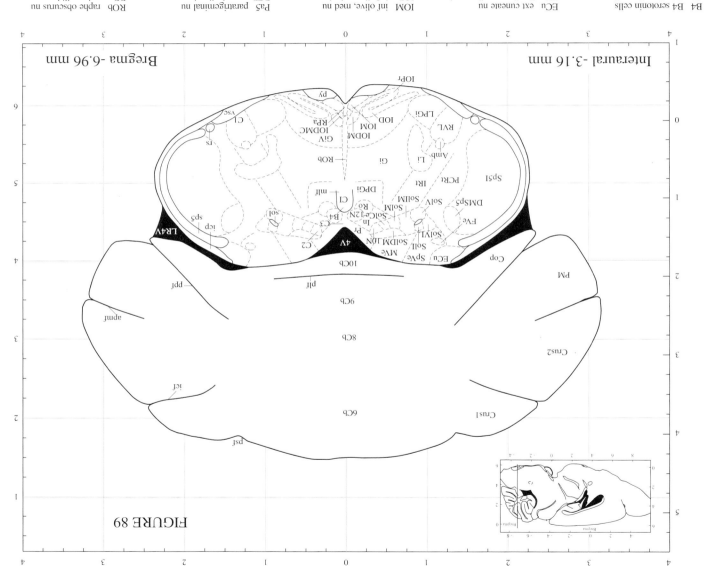

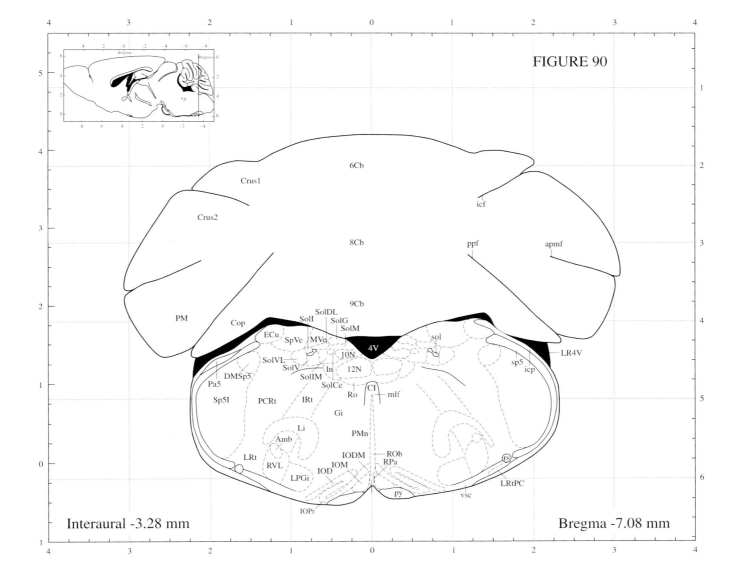

FIGURE 90

Interaural -3.28 mm Bregma -7.08 mm

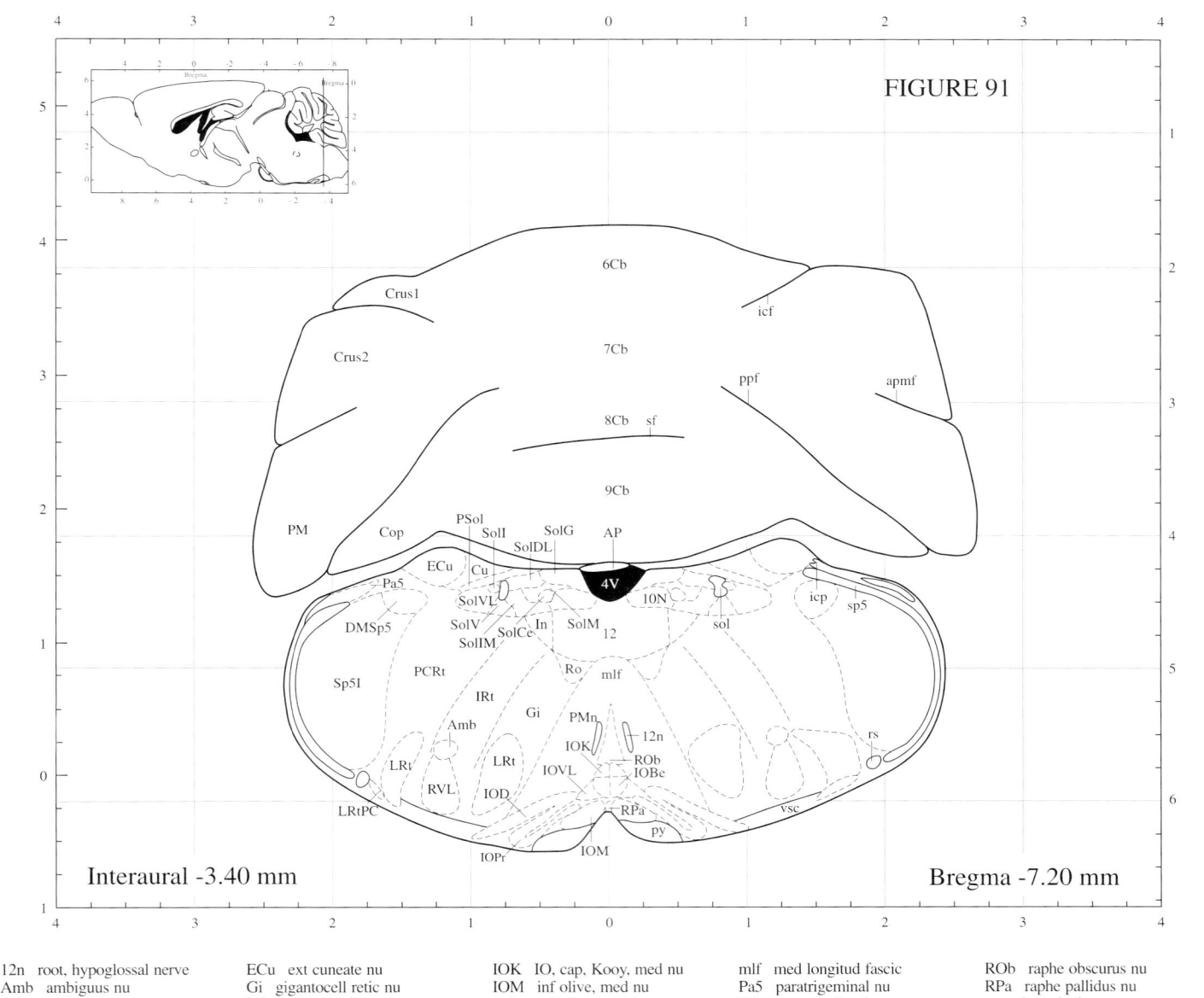

FIGURE 91

Interaural -3.40 mm

Bregma -7.20 mm

FIGURES 91 and 92

4V 4th ventric	12n root, hypoglossal nerve	ECu ext cuneate nu	IOK IO, cap, Kooy, med nu	mlf med longitud fascic	ROb raphe obscurus nu	SolI solitary nu, interstit
6Cb 6th Cerebell lobule	Amb ambiguus nu	Gi gigantocell retic nu	IOM inf olive, med nu	Pa5 paratrigeminal nu	RPa raphe pallidus nu	SolIM solitary nu, intmed
7Cb 7th Cerebell lobule	AP area postrema	icf intercrural fissure	IOPr inf olive, princip nu	PCRt parvicell retic nu	rs rubrospinal tr	SolM solitary nu, med
8Cb 8th Cerebell lobule	apmf ansoparamed fiss	icp inferior cerebell pedunc	IOVL IO, ventlat protrusion	PM paramedian lobule	RVL rostroventlat retic nu	SolV sol nu, vent pt
9Cb 9th Cerebell lobule	Cop copula, pyramis	In intercalated nu, medulla	IRt intermed retic nu	PMn paramed retic nu	sf 2nd fissure	SolVL solitary nu, ventlat
10N dors motor nu, vagus	Crus1 crus 1, ansiform lob	IOB IO, B, med nu	LRt lat retic nu	ppf prepyramidal fissure	sol solitary tr	sp5 spinal trigeminal tr
11n accessory nerve	Crus2 crus 2, ansiform lob	IOBe inf olive, beta	LRtPC lat retic nu, parvicell	PSol parasolitary nu	SolCe solitary nu, ce	Sp5I spin 5 nu, interpolar
12N hypoglossal nu	Cu cuneate nu	IOC IO, C, med nu	MdD medull retic nu, dors	py pyramidal tr	SolDL sol nu, dorsolat pt	vsc vent spinocerebell tr
	DMSp5 dorsmed spin 5 nu	IOD inf olive, dors nu	MdV medull retic nu, vent	Ro nu, Roller	SolG solitary nu, gelati	

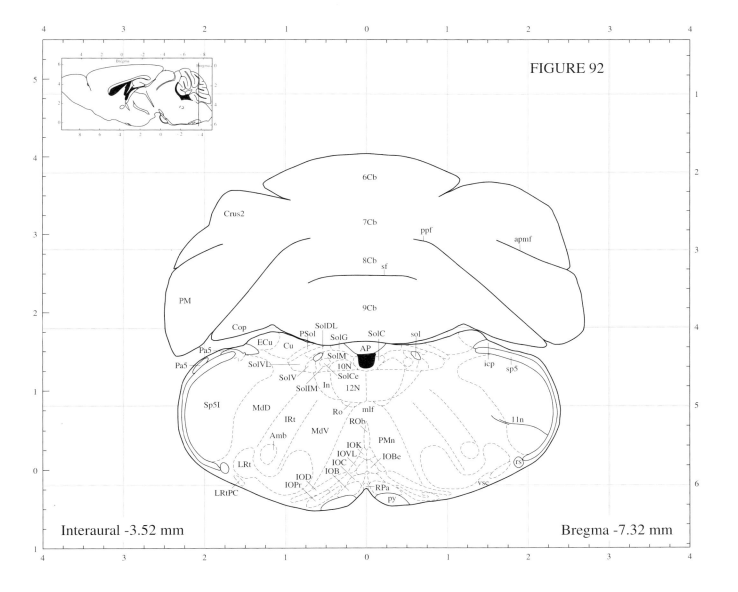

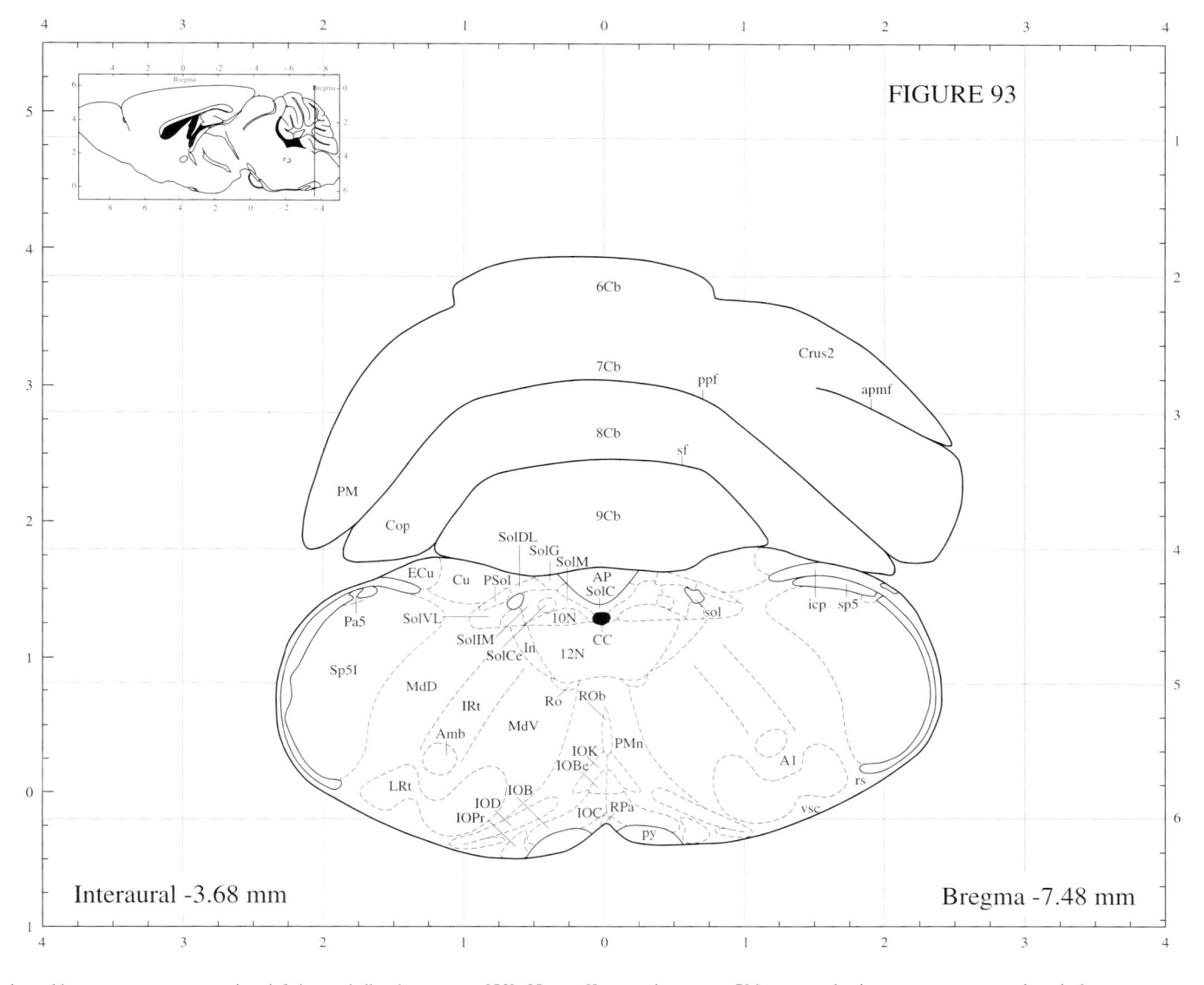

FIGURE 93

Interaural -3.68 mm

Bregma -7.48 mm

FIGURES 93 and 94

6Cb 6th Cerebell lobule	Amb ambiguus nu	icp inferior cerebell pedunc	IOK IO, cap, Kooy, med nu	PMn paramed retic nu	rs rubrospinal tr	SolIM nu solitary tr, intmed
7Cb 7th Cerebell lobule	AP area postrema	In intercalated nu, medulla	IOPr inf olive, princip nu	ppf prepyramidal fissure	sf 2nd fissure	SolM solitary nu, med
8Cb 8th Cerebell lobule	apmf ansoparamed fiss	InM intermedius nu, medulla	IRt intermed retic nu	psf post superior fissure	sol solitary tr	SolV sol nu, vent pt
9Cb 9th Cerebell lobule	CC central canal	IOA IO, A, med nu	LRt lat retic nu	PSol parasolitary nu	SolC solitary nu, comm	SolVL solitary nu, ventlat
10N dors motor nu, vagus	Cop copula, pyramis	IOB IO, B, med nu	MdD medull retic nu, dors	py pyramidal tr	SolCe solitary nu, ce	sp5 spinal trigeminal tr
12N hypoglossal nu	Crus2 crus 2, ansiform lob	IOBe inf olive, beta	MdV medull retic nu, vent	Ro nu, Roller	SolDL sol nu, dorsolat pt	Sp5C spin 5 nu, caud pt
A1 A1 noradrenaline cells	Cu cuneate nu	IOC IO, C, med nu	Pa5 paratrigeminal nu	ROb raphe obscurus nu	SolG solitary nu, gelati	Sp5I spin 5 nu, interpolar
	ECu ext cuneate nu	IOD inf olive, dors nu	PM paramedian lobule	RPa raphe pallidus nu	SolI solitary nu, interstit	vsc vent spinocerebell tr

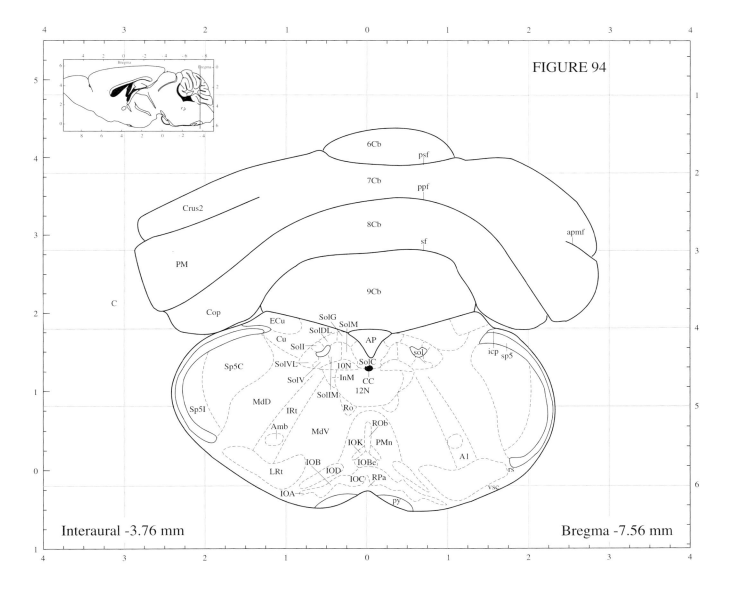

FIGURE 94

Interaural -3.76 mm Bregma -7.56 mm

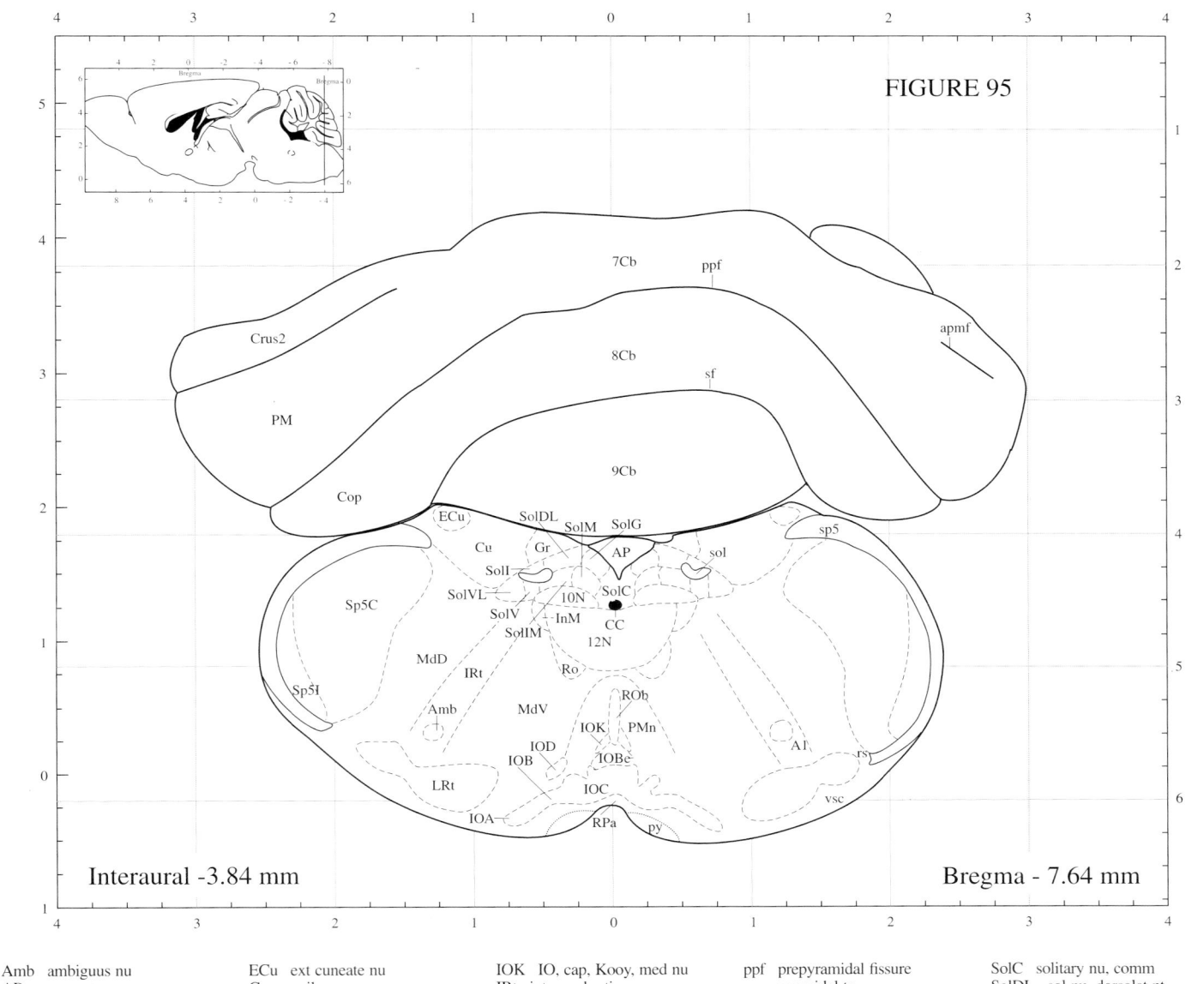

FIGURE 95

Interaural -3.84 mm

Bregma - 7.64 mm

FIGURES 95 and 96

7Cb 7th Cerebell lobule
8Cb 8th Cerebell lobule
9Cb 9th Cerebell lobule
10N dors motor nu, vagus
12N hypoglossal nu
12n root, hypoglossal nerve
A1 A1 noradrenaline cells

Amb ambiguus nu
AP area postrema
apmf ansoparamed fiss
CC central canal
Cop copula, pyramis
Crus2 crus 2, ansiform lob
Cu cuneate nu
cu cuneate fasciculus

ECu ext cuneate nu
Gr gracile nu
InM intermedius nu, medulla
IOA IO, A, med nu
IOB IO, B, med nu
IOBe inf olive, beta
IOC IO, C, med nu
IOD inf olive, dors nu

IOK IO, cap, Kooy, med nu
IRt intermed retic nu
LRt lat retic nu
MdD medull retic nu, dors
MdV medull retic nu, vent
mlf med longitud fascic
PM paramedian lobule
PMn paramed retic nu

ppf prepyramidal fissure
py pyramidal tr
Ro nu, Roller
ROb raphe obscurus nu
RPa raphe pallidus nu
rs rubrospinal tr
sf 2nd fissure
sol solitary tr

SolC solitary nu, comm
SolDL sol nu, dorsolat pt
SolG solitary nu, gelati
SolI solitary nu, interstit
SolIM solitary nu, intmed
SolM solitary nu, med
SolV sol nu, vent pt

SolVL solitary nu, ventlat
sp5 spinal trigeminal tr
Sp5C spin 5 nu, caud pt
Sp5I spin 5 nu, interpolar
ts tectospinal tr
vert vertebral artery
vsc vent spinocerebell tr

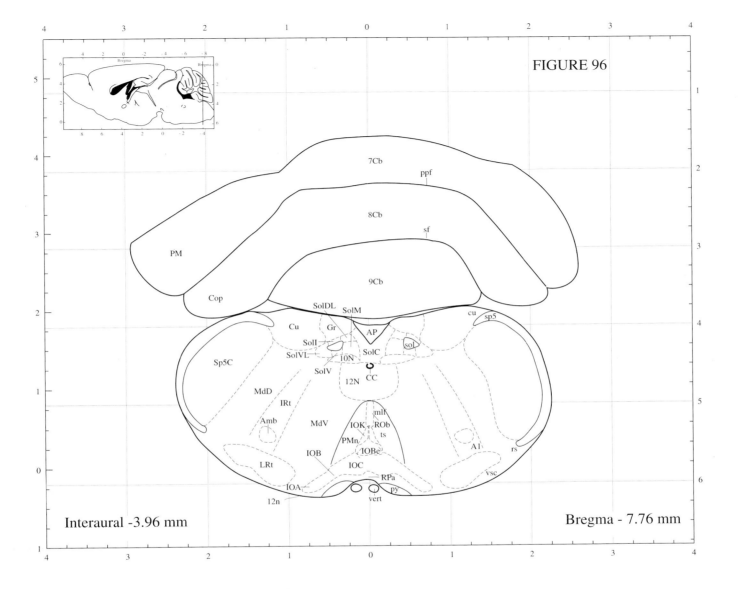

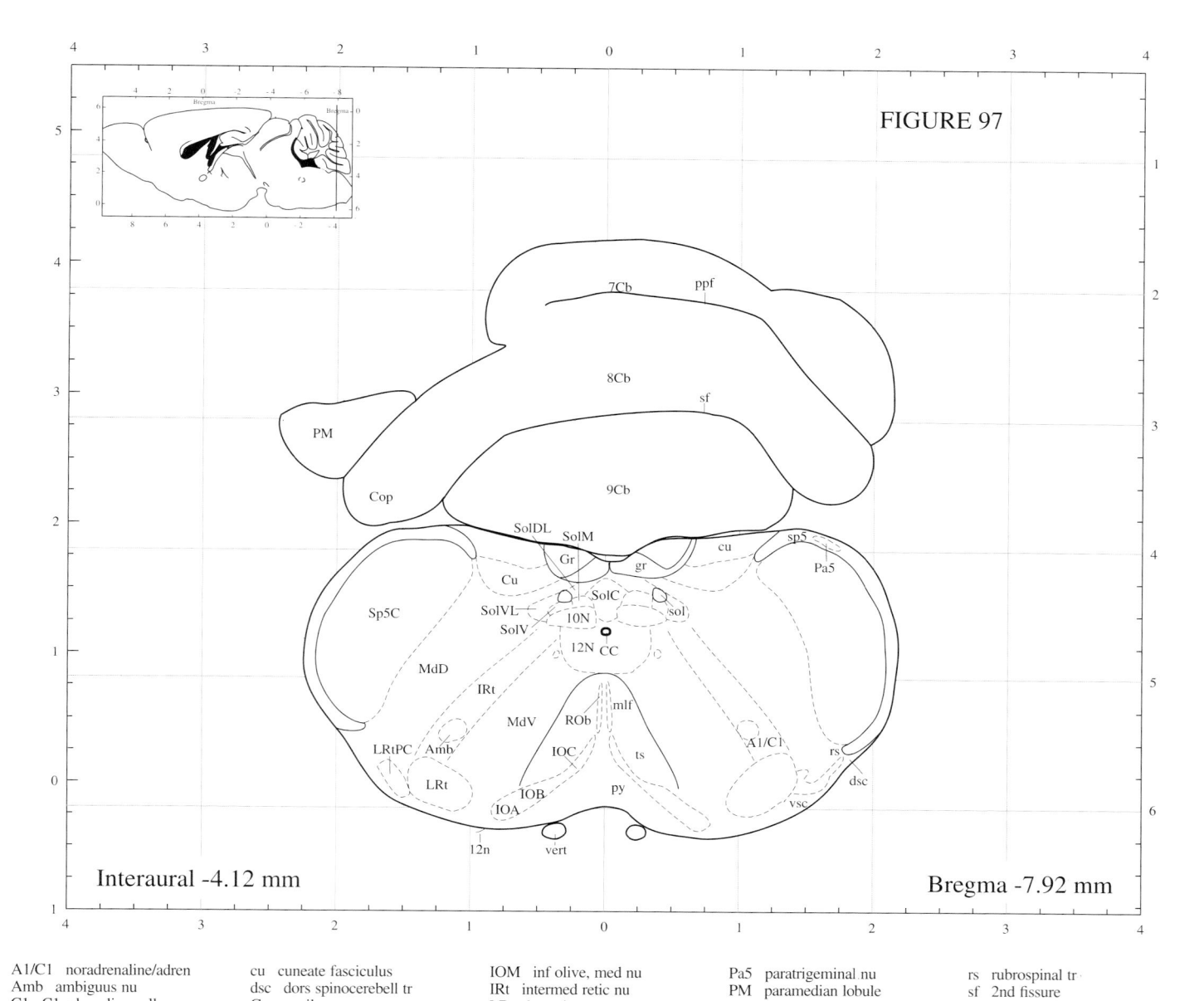

FIGURE 97

Interaural -4.12 mm

Bregma -7.92 mm

FIGURES 97 and 98

7Cb 7th Cerebell lobule
8Cb 8th Cerebell lobule
9Cb 9th Cerebell lobule
10N dors motor nu, vagus
12N hypoglossal nu
12n root, hypoglossal nerve

A1/C1 noradrenaline/adren
Amb ambiguus nu
C1 C1 adrenaline cells
CC central canal
CeCv central cervical nu
Cop copula, pyramis
Cu cuneate nu

cu cuneate fasciculus
dsc dors spinocerebell tr
Gr gracile nu
gr gracile fasciculus
IOA IO, A, med nu
IOB IO, B, med nu
IOC IO, C, med nu

IOM inf olive, med nu
IRt intermed retic nu
LRt lat retic nu
LRtPC lat retic nu, parvicell
MdD medull retic nu, dors
MdV medull retic nu, vent
mlf med longitud fascic

Pa5 paratrigeminal nu
PM paramedian lobule
ppf prepyramidal fissure
py pyramidal tr
pyx pyramidal decussation
RAmb retroambiguus nu
ROb raphe obscurus nu

rs rubrospinal tr
sf 2nd fissure
sol solitary tr
SolC solitary nu, comm
SolDL sol nu, dorsolat pt
SolM solitary nu, med
SolV sol nu, vent pt

SolVL solitary nu, ventlat
sp5 spinal trigeminal tr
Sp5C spin 5 nu, caud pt
ts tectospinal tr
vert vertebral artery
vsc vent spinocerebell

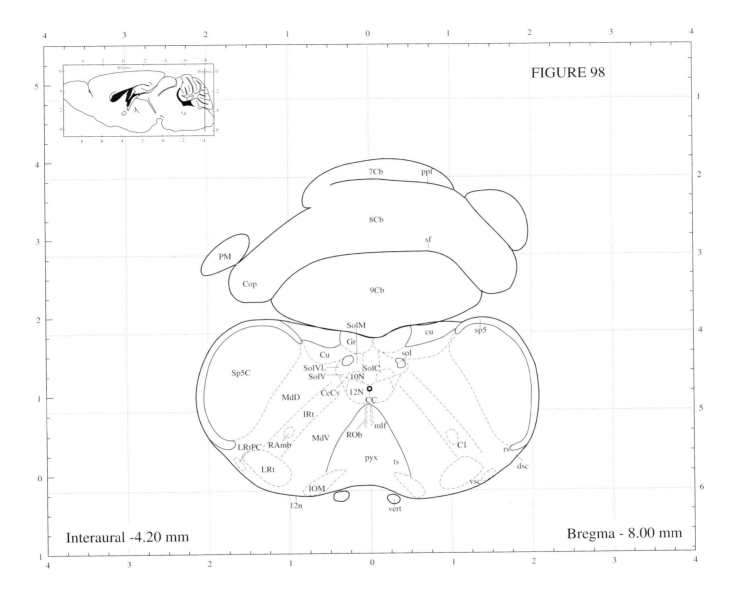

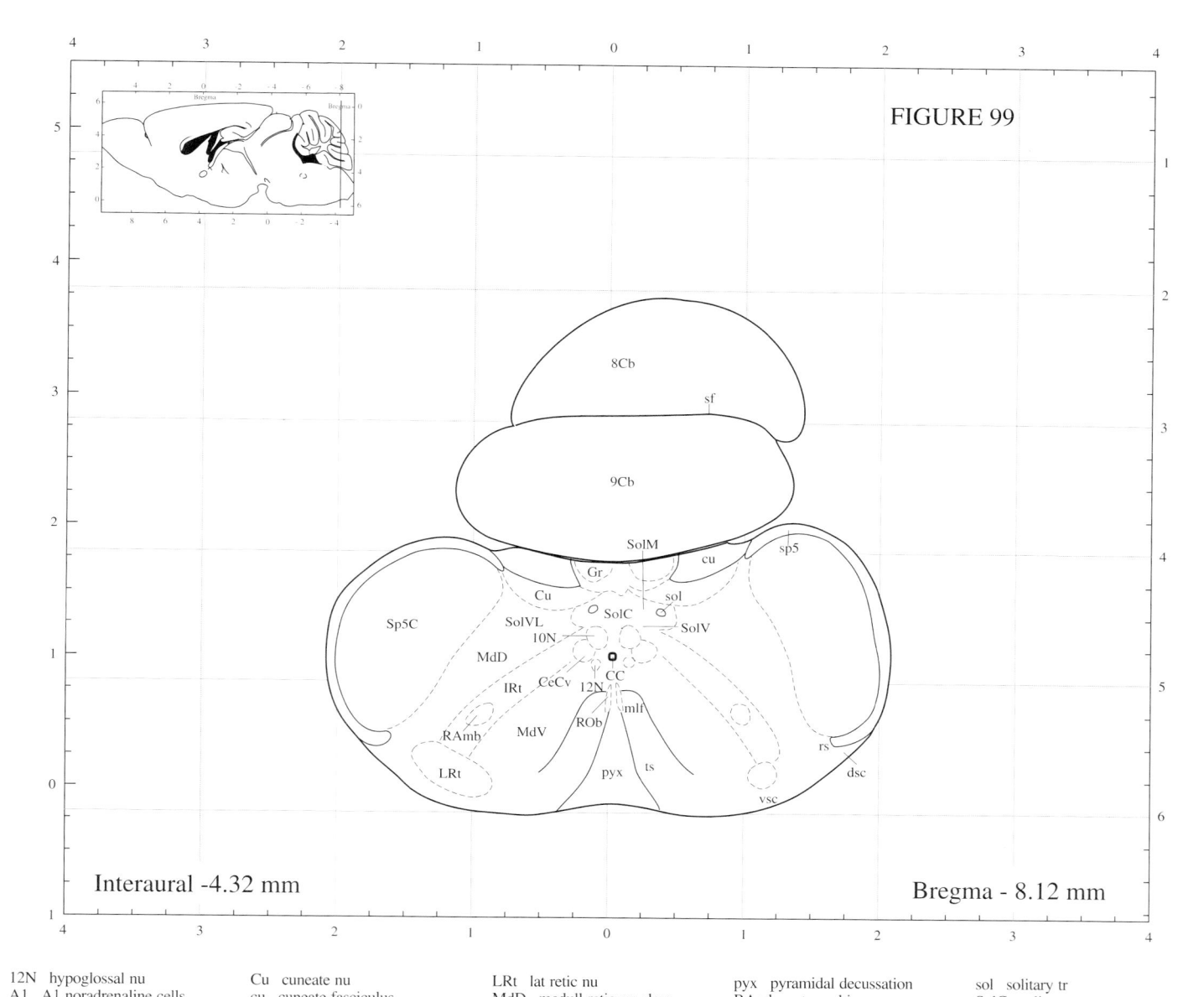

FIGURE 99

Interaural -4.32 mm

Bregma - 8.12 mm

FIGURES 99 and 100

8Cb 8th Cerebell lobule
9Cb 9th Cerebell lobule
10N dors motor nu, vagus
11N accessory nerve nu

12N hypoglossal nu
A1 A1 noradrenaline cells
A2 A2 noradrenaline cells
CC central canal
CeCv central cervical nu

Cu cuneate nu
cu cuneate fasciculus
dsc dors spinocerebell tr
Gr gracile nu
IRt intermed retic nu

LRt lat retic nu
MdD medull retic nu, dors
MdV medull retic nu, vent
mlf med longitud fascic
MnA median acc nu, medulla

pyx pyramidal decussation
RAmb retroambiguus nu
ROb raphe obscurus nu
rs rubrospinal tr
sf 2nd fissure

sol solitary tr
SolC solitary nu, comm
SolM solitary nu, med
SolV sol nu, vent pt
SolVL solitary nu, ventlat

sp5 spinal trigeminal tr
Sp5C spin 5 nu, caud pt
ts tectospinal tr
vsc vent spinocerebell tr

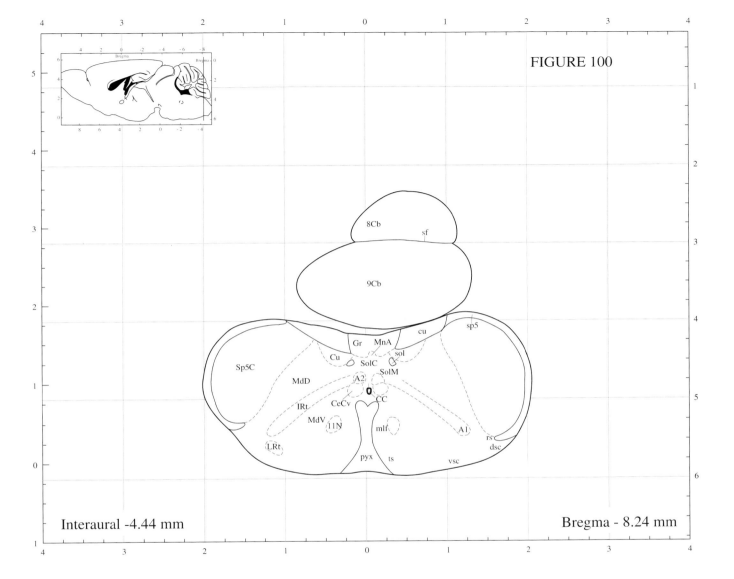